REPRODUCTION IN
BUFFALO

REPRODUCTION IN
BUFFALO

Natural and Assisted Reproductive Techniques

SYED MOHMAD SHAH and
MANMOHAN SINGH CHAUHAN

Notion Press

Old No. 38, New No. 6
McNichols Road, Chetpet
Chennai - 600 031

First Published by Notion Press 2017
Copyright © Syed Mohmad Shah and Manmohan Singh Chauhan 2017
All Rights Reserved.

ISBN 978-1-946556-43-1

This book is dedicated to all the students, researchers and scholars of Reproductive Biotechnology

Table of Contents

Foreword

The aim of any research is to understand the nature and then exploit it for the betterment of humankind and while sustaining the natural resources. This aim is vivid and most expressivi: in every applied scientific research. Animal production and management assumes much significance in the developing world, wherein several smallholders directly depends upon the livestock for food and nutritional security. Among the livestock, the buffalo is an integral and often essential paft of many mixed farming systems, at least in Asian countries, however, in spite of its greater significant, buffalo production has received comparatively lesser attention owing to inherent problems in reproduction. Reproduction ability of the animal is the backbone of any production system and thus understanding the complex process of buffalo reproduction and modulating it towards betterment has long been the interest of researchers. Thus, the book "Reproduction in buffalo" must be a worthy and valuable for animal scientists in general and reproductive biotechnologists in particular. lt gives me an immense pleasure to write a foreword for this book, since this book addresses several basic and applied aspects of the buffalo reproduction besides current advances in reproductive technologies. The book thoroughly discusses buffalo reproductive physiology as well as the assisted reproductive techniques, which provide means for overcoming the intrinsic and natural limitations. Writing a book on a single species, and that too on which meager knowledge is available, is a daunting task. I believe that the book would be much liked by the students because it provides a comprehensive knowledge on all the aspects of the buffalo reproduction. Further, the book would also be highly useful to reproductive biotechnolog ists and researchers.

I congratulate the authors for making such efforts and bringing the book in a clear and simple language and wish them all the best for the success of this book.

A. K. Srivastava

Preface

Buffalo is an indispensable livestock resource that plays a significant role in economy of several countries including India. In view of the key role played in different farming systems by providing milk, meat, draft power and hide, interest in buffalo production is constantly increasing in recent times. Moreover, buffaloes are well-adapted to harsh environments and are capable of utilizing low quality roughages especially agricultural crop residues and by-products. Nonetheless, the inherent low reproductive efficiency of buffalo, compounded further by poor management, remains a major economic problem. Although much advancement has been made in management of reproductive problems in farm animals, serious losses due to sub-fertility or infertility in buffalo remains a major economic problem in the country.

Improvement in the reproductive efficiency of buffalo requires an understanding of the buffalo reproductive system, limitations of the farming system and development of simple intervention strategies along with application of recent Assisted Reproductive Technologies (ART). Specialized reproductive technologies such as artificial insemination, embryo transfer, reproductive cloning and stem cell technology provide new opportunities for faster dissemination of the elite stock. The genetic improvement could be hastened by integrated breeding programs along with application of such advanced biotechnological procedures.

We believe this book would be useful to stimulate new ideas and contribute to better understanding of recent reproductive biotechniques among the students and researchers. With this in mind, we are sure that this book is envisioned as a bridge for researchers and as a tool for technicians and students to facilitate them in developing new research concepts and facilities at their work places.

Authors

Acknowledgements

It was in August 2010, the time when I consciously registered for my PhD at National Dairy Research Institue's much cherished and widely famous Animal Biotechnology Center, that a curiosity to write a book on Reproduction in Buffalo developed in me. All my colleagues and juniors were in distress and anxiety to grab a book that would suffice us with the knowledge about buffalo Reproduction- natural and assisted, for this was one among our main subjects. We all felt dismayed for finding nothing much useful anywhere. We turned to seniors for help but in vain. It was then, in those sultry days, that I decided to prepare a book- a book that would at least provide a firsthand knowledge of reproduction in general, and of bubaline reproduction in particular, for it was the research mandate of my institute. My aim was to aid the students and scholars for preparing for their exams and the almost weekly presentations. I aimed to write a book which would be useful to teachers and researchers as well. Though my primary target is the Beginner in the subject, I hope the experts too won't get bored.

My foremost and most important THANKS would be to the Reproductive biologists, especially of my Institute for daring to undertake such a daunting and challenging subject, and for spending their time, energy and neurons for studying the mysterious phenomenon of Reproduction.

My candid and merry THANKS to my guide and co-author of this book, for everything he did and not-did for me. I will always remember him for his encouragements, advices and above-all having a deep faith in me.

I would love to thank my friends, colleagues and all those researchers from whose knowledge I have benefitted from in writing this book.

My special thanks to my full coterie of friends who have always been there to accompany me, to love me, to care for me, to help me out of a problem, to throw me into a disaster, to criticize me, to fight with me, to fight on my behalf, to empty my wallet, to be my credit cards, to embolden me, to make me feel feeble at times and above all to laugh and weep with me. I thank you all for all you did and are doing for me!

I humbly offer my thanks to my family for providing me with an amicable and supportive home. My special and loving THANKS to my sister for sipping me infinite tea cups, during the draft preparation.

At last, but with all the humility, I submit my THANKS to Almighty for giving me yet another chance to serve, though minutely, my profession.

I pray this book be useful to students, scholars and all the concerned for that is the only purpose it has been prepared for!

Sincerely

Syed M Shah

Abbreviations and Symbols

%	:	Percent
°C	:	Degree Celsius
mg	:	Microgram
μM	:	Micromolar
ml	:	Microlitre
AI	:	Artificial insemination
AR	:	Acrosome reaction
ART	:	Assisted Reproductive techniques
BO	:	Brackett and Oliphant
BSA	:	Bovine serum albumin
bFGF-2	:	Basic fibroblast growth factor 2
BLIMP1	:	B Lymphocyte Induced Maturation Protein 1
BME	:	Beta mercaptoethanol
BMP4	:	Bone morphogenetic factor 4
buFF	:	Buffalo follicular fluid
buESC	:	Buffalo embryonic stem cells
CO_2	:	Carbon dioxide
COC	:	Cumulus-oocyte-complex
CCM	:	Cumulus conditioned medium
COH	:	Controlled ovarian hyperstimulation
cm	:	Centimeter
DEPC	:	Diethyl pyrocarbonate
DMEM	:	Dulbecco's modified eagle's medium
DMSO	:	Dimethyl sulphoxide
DPBS^{++}	:	Dulbecco's phosphate buffered saline with Ca^{2+} and Mg^{2+}
DPBS^{--}	:	Dulbecco's phosphate buffered saline without Ca^{2+} and Mg^{2+}
dpc	:	Days post coitus
DPPA	:	Developmental pluripotency associated
EAA	:	Essential amino acids
EB	:	Embryoid body
EC	:	Embryonal carcinoma

ECM	:	Extra cellular matrix
EDTA	:	Ethylenediamine tetra-acetic acid
EG	:	Ethylene glycol
EGF	:	Epidermal growth factor
ELISA	:	Enzyme linked immunosorbent essay
ERK	:	Extracellular signal receptor kinases
ES cells	:	Embryonic stem cells
FBS	:	Fetal bovine serum
FOXD3	:	Forkhead box D3
FSH-p	:	Follicle stimulating hormone-porcine
g	:	gram
GAPDH	:	Glyceraldehyde 3-phosphate dehydrogenase
GDF	:	Growth and differentiation factor
GIFT	:	Gamete intrafallopian transfer
GnRH	:	Gonadotrophin releasing hormone
GP130	:	Glycoprotein 130
GSC	:	Germ stem cell
GV	:	Germinal vesicle
GVBD	:	Germinal vesicle breakdown
HAM	:	Hyperactivated motility
HD	:	Hanging drop
hEFs	:	Human embryonic fibroblasts
HEPES	:	(4-(2-hydroxyethyl)-1-piperazineethanesulfonic acid)
hES cell	:	Human embryonic stem cell
HGC	:	Hand guided cloning
hpi	:	hours post insemination
ICSI	:	Intra cytoplasmic sperm injection
IGF	:	Insulin-like growth factor
IL	:	Interleukein
IVC	:	*In vitro* culture
IVF	:	*In vitro* fertilization
IVM	:	*In vitro* maturation
IVP	:	*In vitro* production
I.U.	:	International Unit
IVEP	:	*In vitro* embryo production
JAK	:	Janus family kinase
KO-DMEM	:	Knockout Dulbecco's modified Eagle's medium
KoSR	:	Knockout serum replacement

LH	:	Luteinizing hormone
MAPK	:	Mitogen activated protein kinase
mCR2aa	:	Modified Charles-Rosenkrans+ amino acids- 2
MEFs	:	Mouse embryonic fibroblasts
mES cell	:	Murine embryonic stem cell
mLIF	:	Murine leukemia inhibitory factor
MIS	:	Meiosis inducing substance
MPF	:	Maturation promoting factor
MPS	:	Meiosis-preventing substance
MOET	:	Multiple ovulation and embryo transfer
mg	:	Miligram
MII	:	Metaphase II
ml	:	Mililitre
mM	:	Millimole
mm	:	Millimeter
NEAA	:	Non-essential amino acids
NT	:	Nucleus transfer
NTES	:	Nucleus transfer embryonic stem
OCT-4	:	Octamer binding transcription factor
OLS	:	Oocyte-like structure
O_2	:	Oxygen
OPU	:	Ovum pick up
OSP	:	Osteopontin
PBS	:	Phosphate buffer saline
PCFR	:	Primary colony formation rate
PCR	:	Polymerase chain reaction
PGC	:	Primordinal germ cell
PHA	:	Phytohemagglutinin
PLC	:	Phospholipase C
PMSG	:	Pregnant mare serum gonadotrophin
PRM	:	Protamine
PRMT	:	Protein arginine methyltransferase
PVA	:	Polyvinyl alcohol
PVDF	:	Polyvinylidene fluoride
RA	:	Retinoic acid
RARE	:	Retinoic acid responsive element
REX-1	:	Reduced expression 1
RIA	:	Radio immunoassay

ROCK	:	Rho-associated coiled kinase
RTKs	:	Receptor tyrosine kinases
RT-PCR	:	Reverse transcriptase- polymérase chain reaction
SCM	:	Syed-Chauhan medium
SCNT	:	Somatic cell nuclear transfer
SDF	:	Stromal cell-derived factor
SCF	:	Stem cell factor
SDS-PAGE	:	Sodium dodecyl sulfate polyacrylamide gel electrophoresis
SLBP	:	Stem loop binding factor
SOX2	:	Sex determining region Y gene - related HMG box 2
SSEAs	:	Stage specific embryonic antigens
SSC	:	Static suspension culture
STAT3	:	Signal transducers of activation
STF	:	Seminiferous tubule fluid
SYCP	:	Synaptonemal complex protein
T0	:	TCM-199 without FBS
T2	:	TCM-199 supplemented with 2% FBS
T10	:	TCM-199 supplemented with 10% FBS
T20	:	TCM-199 supplemented with 20% FBS
TAE	:	Tris acetate EDTA
TBST	:	Tris-buffered saline with Tween 20
TCM-199	:	Tissue culture medium-199
TE	:	Trophectoderm
TEMED	:	N,N,N,'N'-tetramethyl- ethane-1,2-diamine
TGF-β	:	Transforming growth factor beta
TNP	:	Transition protein
TRAs	:	Tumor rejection antigens
TVOR	:	Transvaginal oocyte retrieval
WB	:	Western blotting
WNT	:	Wingless signaling pathway
ZIFT	:	Zygote intra fallopian transfer
ZP	:	Zona pellucida

Chapter 1

Buffalo – The Black Gold of India

Introduction

Livestock has been an integral part of human civilization right from the nomadic era to this industrialized age, and is believed to remain the renewable battery in the post-industrial era also. It has been a source of food, means of tillage and transport and a reliable and gentle companion to humans throughout the ages. Most of the farm animal species were widely commercialized from traditionally wild and natural habitats in early centuries with the exception of buffalo whose commercial-size operation became pronounced a bit later with the introduction of domesticated breeds outside of Asia. Currently more human beings depend on domestic water buffalo for milk, meat and draught power than on any other domestic animal.

Scientific Classification of buffalo

Kingdom	Animalia
Phylum	Chordata
Class	Mammalia
Order	Artiodactyla
Suborder	Ruminatia
Family	Bovidae
Subfamily	Bovinae
Genus	*Bubalus*
Species	Bubalis
Binomial Name	*Bubalus bubalis*

Carl Linnaeus first described the genus Bos and water buffalo under the binomial name *Bubalus bubalis* in 1758. The classification, however, is uncertain with some authorities treating the wild and domestic forms as conspecific, while others treat them as different species, thereby creating an inconsistency in their nomenclature. To achieve consistency in naming the wild and domesticated forms, International Commission on Zoological nomenclature (2003) ruled scientific name *Bubalus arnee* for the wild form while the domestic form was named as *Bubalus bubalis*.

Domestication and Distribution

The Arni or wild buffalo survives in wild in India, Pakistan, Bangladesh, Nepal, Bhutan, China, Vietnam, Taiwan, Philippines, Indonesia and Thailand. The domestic form is predominant in Asia, South America, North Africa, Europe and North Australia. In Asia, the population of wild water buffalo has become sparse and it is feared that no pure bred wild water buffalo exists as of now.

Water buffaloes were domesticated in India and China, respectively, about 5000 and 4000 years ago. They were introduced into Europe from India or other oriental countries and to Italy in the reign of King Agilulf. European buffaloes are considered to be the same breed as Mediterranean buffalo. In Italy, the Mediterranean type was particularly selected and is named as the Mediterranean Italian breed to distinguish it from other European breeds from which it differs genetically. Water buffaloes were introduced into Australia (Darwin and Arnhem land) in 1880s from India. They became feral (returned back to wild state) causing significant environmental damage. Indonesia has both riverine and swamp buffalo populations and is the only country in the world where *Bubalus depressicornis* (Anoa) still exists. It is regarded as the smallest bovid in the world with a height of 80 cm, live weight of 200 kg and 30 cm long horns. The indigenous buffalo of Africa is regarded as a member of another species and genus (*Syncerus caffer*). They possess 52 chromosomes in comparison to 50 in *Bubalus* genus of Asia and Europe.

Based on morphological and behavioural criteria, two types of water buffalo are generally recognized - the **river type** (*Bubalus bubalis bubalis*) found in Indian subcontinent, Balkans and Italy (South Asia) and the **swamp or carabao type** (*Bubalus babalis carabanesis*) found in Assam, Philippines to Yangtze valley of China (South-east Asia). The swamp buffalo has 48 chromosomes while the riverine has 50. The two types do not readily interbreed, though fertile offspring occur between the two. The offspring has 49 chromosomes and sometimes possess the best traits of the two. The present day river buffalo is the result of complex domestication process involving more than one maternal lineage and a significant maternal gene flow from wild populations after initial domestication events.

World water buffalo population is about 172 million and an approximate 96% of this population is found in Asia, the native home. Among Asian countries, India possesses maximum share of this population (57%) harboring primarily the 10 well-defined river type breeds, while swamp type occur in small areas in north-eastern part of the country and are not distinguished into breeds. China, Pakistan, Philippines, Vietnam, Bangladesh and Sri Lanka possess an appreciable population share. Mediterranean buffaloes are found in Europe, Italy, Romania, Bulgaria, Greece, Albania, Kosovo and Republic of Macedonia. In Bulgaria, they were crossbred with Indian Murrah breed resulting to Bulgarian Murrah. Populations in Turkey are of Anatolian breed. Some breeds like Mediterranean (from Italy), Murrah and Jafarabadi (from India) and Carabao (from Phillippines) were introduced into South America (Amazon river basin) in 1895.

The estimated population of wild buffalo stands currently at 4000. This number includes all wild population together with feral herds and hybrid buffaloes. The Indian wild buffalo is found in India, Pakistan, Bangladesh, Nepal, Bhutan and Thailand. This buffalo has the largest horns of any living animal with an average spread of 1m. The buffalo is listed as endangered in the IUCN (International Union for Conservation of Nature) Red List since 1986 and is included in CITES (Convention on International Trade in Endangered Species) under Appendix III in Nepal. It is also legally protected in Bhutan, India, Nepal and Thailand.

Body features and behaviour

The skin of river buffalo is black but some species may have dark slate-coloured skin. Swamp buffalo have grey skin at birth which becomes slate blue later. Riverine buffalo have comparatively longer faces, smaller girth and bigger limbs than swamp buffalo whose face is shorter with wide muzzle and prominent eyes. Swamp buffalo is heavy-bodied and stockily built with short body and large belly. The dorsal ridge extends backwards and tapers off gradually in river type buffalo, while it ends abruptly just before the end of chest in swamp buffalo. Horns grow downward and backward then curve upward in a spiral in riverine type, while the curve is semi-circular and more or less in the plane of forehead in swamp type. Tail is short in both types, reaching only to the hocks. Height at withers is 129–133 cm in males while females are comparatively shorter (120–127 cm). The body weight ranges from 300–600 kg for the domestic breed, while wild type grows larger (900–1200 kg). Wild buffalo also lacks round belly found in domesticated form. The rumen of water buffalo contains a larger population of bacteria, particularly the cellulolytic bacteria and fungal zoospores, while protozoa count is lesser than that of cattle. Rumen ammonia nitrogen and pH are also higher than cattle.

Riverine buffalo prefer deep clear water, while swamp buffalo wallow in marshy waters and mud holes which they make with their horns. Buffalo are well adapted to hot and humid climate to a temperature range of 0 to 30°C. Swamp buffalo are mainly used as draught animals and very few are reared for milk as they produce 1 to 2 liters of milk per day. Riverine buffalo, in contrast, are raised mainly for milk production.

Reproduction

Riverine buffalo generally attain sexual maturity at a later age than swamp buffalo. Age at first estrous of heifers varies between breeds, ranging from 13–33 months. Mating at first estrus is often infertile, until up to 3 years. Males attain sexual maturity at about 3–3.5 years of age. Despite being polyestrous, reproductive efficiency shows wide variation throughout the year, with display of a distinct seasonal change in estrus, conception and calving rates. Gestation period in riverine buffalo ranges between 300–320 days, while swamp buffaloes carry their calves for one to two weeks longer.

Significance

Buffalo are the second largest source of milk supply in the world. Dairy cattle produce 84% of total milk in the world, with an average fat and protein content of 4 and 3.5%, respectively, which is lesser than that of buffalo milk (7% and 4.5%, respectively). Thus, in terms of energy-corrected milk, it makes a greater food contribution than what the actual volume reflects. Buffalo has been used as draught animal for centuries and is an important beast of burden in Asian farming where it is used for ploughing, tilling, puddling, haul carts, water pumping etc. Their legs can withstand wet conditions better than cattle and their large hoofs and flexible foot joints, makes them an ideal animal for work in deep mud of paddy fields. This advantage earned them the sobriquet of "the living tractor of the East." In India, more than half of the total milk (56–57%) is produced by buffaloes, despite numbering only half to cattle population. In addition to providing protein and fat rich milk to Indian population, buffalo serve as a source of employment and livelihood to millions of landless, marginal and small farmers. Such is its contribution to sustenance and livelihood of Indian farmers that buffalo is eulogized as "the black gold of India."

Water buffalo milk, with higher levels of total solids as compared to cow milk, is ideal for processing into value added dairy products. The higher fat content makes cream churning faster with higher overrun as compared to cow milk. Buffalo butter displays more stability and ghee has a better texture and bigger grain size than that obtained from cow milk. The peroxidase activity in buffalo milk is 2–4 times higher than cow's milk, accounting for its higher natural preservability. Water buffalo meat, carabeef, is a major source of export revenue to India. A buffalo carcass in general has a higher proportion of muscle and a lower ratio of bone and fat than a cattle carcass. The hides provide tough and useful leather, often used for shoes and motor cycle helmets while horns are made into earrings and musical instruments like ney and kaval. Buffalo also has a significant impact on culture and religion. Buffalo is used as a sacrificial animal in Indonesis and China. Hindu lore has it that god of death, Yama, rides on a male water buffalo. Such is the importance of buffalo in Philippines that *carabao* subspecies is considered as a national symbol. Buffalo also serves as an entertainer in various countries of the world. The buffalo fighting festivals of Assam (Moh Juj), Vietnam (Do Son and Hai Luu), Thailand (Ko Samui) and Indonesia (Ma Pasilaga Tedong) are celebrated on special occasions like New Year and commencement of spring season. Buffalo racing festivals are held in Babulang, India, Cambodia, Thailand, Malaysia, etc. Worth mentioning is the fact that farm animals are in the "Buffalo Beauty Pageant," a Miss Farmer beauty contest and a comic buffalo costume contest of Thailand.

Research

India holds two distinctions as its association with buffalo is concerned. 1) It possesses the maximum number of buffalo and ranks first in buffalo milk production; 2) India is the first country in Asia for starting scientific and technological developments in buffalo nutrition, production, reproduction and genetic improvement. The research encompasses all aspects of buffalo husbandry from feed conversion efficiency studies, milk composition analysis, pedigree analysis, breeding programmes, semen/ ova and embryo cryopreservation, artificial insemination, embryo transfer technology, *in vitro* embryo production, cloning, transgenics and stem cell technology. The buffalo research is mainly carried in Indian Council of Agricultural Research (ICAR) institutes, notably National Dairy research Institute (NDRI),

Central Institute of Research on Buffalo (CIRB) and Indian Veterinary Research Institute (IVRI). We, at NDRI, have achieved a distinction for producing first buffalo embryonic stem cell lines as well as the first cloned buffalo calf through indigenously developed Hand-guided cloning technique. Moreover, India has implemented a number of national programmes such as Green revolution (to increase crop production), White revolution (to increase milk productivity) and Red revolution (to increase meat production), particularly with regard to buffalo. Such is the importance of buffalo in developed world that it is considered to be the "Animal of the future" in America. The main character of buffalo that creates such an enthusiasm is its extraordinary ability to convert fibre into energy and superiority in use of tropical forage and agricultural by-products in comparison to cattle, in addition to its rusticity, adaptability to changing climates and high fertility rates than bovines. It is felt that buffalo does not compete with humans for it does not necessarily use the main production from the crops and serves as an efficient tool in recycling of nutrients in integrated production system.

Buffalo breeding in modern husbandry is regarded as a synonym for low production costs and high levels of productivity. In this world of food and feed shortage, buffalo is considered as an "International animal of promise" for such traits as efficient feed conversion and sustenance on agricultural by-products in addition to its rustic behavior and higher adaptability than cattle.

Buffalo Breeds

The general classification of bubaline species is given schematically as under:

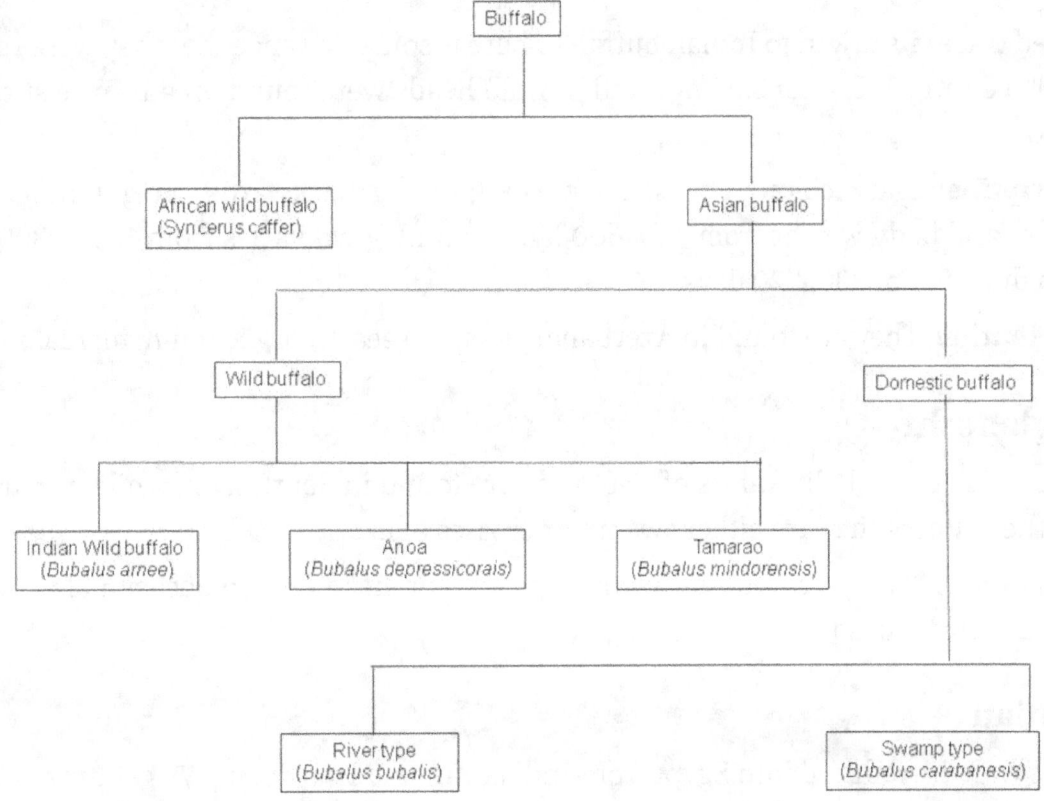

India possesses the best milk breeds of buffalo. Out of the 18 major breeds of buffalo 12 are kept primarily for milk production. Buffalo of Indian sub-continent can be grouped into 5 distinct groups:

A) **Murrah group**: It includes Murrah, Nili-Ravi, Kundi and a new breed Godavari.

B) **Gujrat group**: It includes Surti, Jaffarabadi and Mehsana.

C) **Uttar Pradesh Group**: It comprises of Bhadawari and Tarai.

D) **Central Indian group**: It includes Nagpuri, Pandharpuri, Manda, Jerangi, Kalhandi and Sambalpuri.

E) **South Indian group**: It has Toda and South Kanara.

A brief description of different buffalo breeds of the world is given in the following discussion.

1. Anatolian

This breed has been raised in Turkey and traces its origin to India.

Description: Black colour, long hair and frequent white switch on tail. Height at withers is about 138 cm and body weight ranges from 200–500 kg. Milk yield per lactation period of 220–270 days varies from 700–1,000 kg.

Distribution: Black sea region, Middle Anatolia, Thrace, Hatay, Mus, Kars, Dyarbakir, Afyon and Siwas.

2. Azeri or Caucasian

This breed traces its origin to Indian buffalo. There is some evidence that they were raised in Iran in 9th century B.C. after six engraved buffalo heads were found on a bronze stick from this period.

Description: Black colour and short horns growing backwards. Height ranges from 133–137 cm and body weight from 300–600 kg. The milk yield varies from 1200–1300 kg for lactation duration of 200–220 days.

Distribution: They are found in Azerbaijan, Caspian sea, Georgia and Armenia.

3. Bangladeshi

Indigenous Bangladeshi buffaloes of river type are found in South-west, while in remaining parts of the country they are either swamp or crosses of exotic breeds.

Description: Black in colour with white spot on forehead and an occasional tail-switch. Horns are curled and short.

4. Egyptian

Buffaloes were introduced into Egypt from India, Iran and Iraq during 7th century.

Description: Black-grey in colour with sloping rump and low tail. Horns are lyre to sword-shaped while the head is long and narrow. Height varies from 144–178 cm and body weight from 500–600 kg. Milk yield is from 1200–2100 kg for a lactation period of 210–280 days.

Distribution: All over the country, mainly in peri-urban areas and Nile delta.

5. Lime

The pure Lime breed has originated from wild Arna and has been domesticated throughout the known history of Nepal.

Description: Light brown body colour with chevrons of grey or white hair below the jaws and around the brisket. The horns are small sickle-shaped and curved towards the neck. Height is about 115 cm and body weight about 399 kg. The average milk yield is 875 kg for a lactation length of 351 days.

Distribution: The breed is found in high hills and hill river valleys in Nepal.

6. Mediterranean or European

It has originated from Indian buffalo and was introduced into Europe with the advent of Islam in 6th and 7th centuries.

Description: Coat colour is black, black and brown or dark grey. Horns are flat at the bottom and pointed slightly outwards while the top is pointed inwards. The milk yield varies from 900–4000 kg for a lactation length of 270 days.

Distribution: Italy, Romania, Brazil, Greece, Albania, Macedonia, Germnay, Netherlands, Switzerland, United Kingdom and Hungary.

7. Parkote

These are typical buffalo of mid-hills and river valleys in Nepal. Their population, however, is declining due to cross-breeding with Lime and Murrah buffaloes.

Description: Dark coat colour, medium built body size with sword-shaped horns directed laterally or towards the back. Height is about 114 cm and body weight approximates 410 kg. The lactation duration is 351 days, yielding 875 kg of milk.

Distribution: The breed is raised in mountains, high hills and hill river valleys of Nepal.

8. Khuzestani

It is likely the biggest buffalo breed in the world.

Description: They have short horns which grow upward forming a ring at the end. Height at withers is 141–148 cm and the body weight varies from 600–800 kg. Milk yield is 1300–1400 kg for a lactation period of 200–270 days.

Distribution: Khuzestan and Lorestan provinces of Iran, and Baghdad and Mosul provinces of Iraq.

9. Bulgarian Murrah

This breed arose from up gradation of local buffalo of Bulgaria with Indian Murrah buffalo imported from 1962–1990.

Description: Black or black and brown or dark grey in colour. Adult body weight ranges from 600–700 kg while the milk yield is 1800 kg for a lactation period of 270–305 days.

Distribution: They are found allover Bulgaria, Romania and South America.

10. Murrah

It is the most important and well-known breed of the world.

Description: Black colour with short and tightly curled horns. Height at withers varies from 133–142 cm and body weight from 650–750 kg. Milk yield goes up to 1800 kg for a lactation length of 305 days.

Distribution: It originated in Haryana and spread to Punjab, Ravi and Sutlej valleys, North Sind and Uttar Pradesh. It has been exported to Brazil, Bulgaria and many other countries of eastern Asia.

11. Kundi

Kundi is the second most important breed of Pakistan.

Description: Coat colour is black and has short horns. Height varies from 125–135 cm while body weight from 600–700 kg.

Distribution: It is wide-spread in South Pakistan's Sind region.

12. Nili-Ravi

Nilli and Ravi were two different breeds until 1950 but after this period it was difficult to distinguish between the two breeds probably due to overlapping selection criteria of breeders, thus, popularizing the name Nili-Ravi.

Description: It is similar to Murrah in almost all characteristics except for white-markings on extremities and walled eyes. The horns are less curled than in Murrah. Milk yield is 2000 kg in lactation duration of 305 days.

Distribution: It is the most important livestock in Pakistan and is also present in Indian state of Punjab.

13. Bhadawari

It is an improved local breed of India developed as a result of selection of Indian breeds.

Description: Coat is copper-coloured and the neck presents a typical white colour ring. Tail switch is white or black and white and horns are short and grow backwards. Height varies from 124–128 cm and body weight ranges from 425–475 kg. Lactation length is approximately 270 days and the yield is 760–800 kg.

Distribution: The buffalo is reared in Agra and Etawah districts of Uttar Pradesh and in Bhind and Morena districts of Madhya Pradesh.

14. Tarai

This breed is well adapted to the difficult climatic and feeding conditions of Tarai region.

Description: The coat is black to brown with an occasional white blaze on forehead. Horns are long and flat with coils bending backwards and upwards. Height varies from 120–127 cm and body weight from 325–375 kg. Lactation length is 250 days with a yield of 450 kg.

Distribution: The buffalo is raised in Uttar Pradesh and Madhya Pradesh.

15. Nagpuri

It is an improved local breed as a result of selection of Indian breeds.

Description: Black in colour with occasional white markings on face, legs and switch. Horns are flat-curved and carried back near to the shoulders. Height is 130–140 cm and weight ranges from 408–522 kg. The milk yield is 825 kg for a lactation length of 240 days.

Distribution: The breed is widespread in Nagpur and some districts of Madhya Pradesh.

16. Surti

It is the result of selection of Indian breeds of buffalo and is one of the most important breeds of Gujrat and Rajasthan.

Description: Black coat colour with two white cheverons on the chest. Horns are flat, sickle shaped, directed downward and backward and then turned upwards at the tip to form a hook. Milk yield is 2090 kg for a lactation length of 350 days.

Distribution: The breed is mainly concentrated in Mahi and Sabarmati rivers in Gujrat.

17. Manda/Ganjam

This is also an improved local breed of India. It is a hardy breed and is able to work under the hot sun.

Description: The coat is uni-coloured or ash grey or brown. Some animals are silver-white in colour. Horns are broad and emerge slightly laterally, making half-circles. The buffalo yields about 700 kg of milk in a lactation period of 290 days. Height at withers is approximately 120 cm and the body weight varies from 320–350 kg.

Distribution: The breed is found in Koraput district and adjoining parts of Malkangiri and Newarangpur districts of Orrisa.

18. Mehsana

Mehsana is a locally improved breed of Indian buffalo and is a cross of Surti and Murrah.

Description: Jet black in colour with sickle-shaped horns but more curved than Surti. Body weight ranges from 484–565 kg. The buffalo produces 1800–2700 kg of milk for a lactation period of 305 days.

Distribution: The breed is widespread between Mahi and Sabarmati rivers in Gujrat.

19. Toda

This breed is semi-wild and is raised under semi-nomadic conditions.

Description: It is very attractive animal with slightly reddish coat colour. Its chest is very broad in front, legs are short and tail is long. The height at withers ranges from 150–160 cm and body weight approximates 380 kg. Milk yield is 500 kg for a lactation period of 200 days.

Distribution: The breed is found in the Nilgiri territory of Tamil Nadu and is reared in special families called Mandu.

Further reading

Acharya RM and Bhat PN. (1989). Status paper on buffalo production and health. Proc. FAO round table held in conjunctions with II World Buffalo Congress. India, 12–17 Dec., 1988. pp. 11–37.

Bhat PN, Kumar R and Raheja KL. (1983). Breeding behavior of murrah buffalo. In: Proceedings of the Fourth International Congress of SABRAO. pp. 91–97.

Chaudhary RA and Ahmed W. (1978). Buffalo Breeding of Pakistan and Programmes for their Improvement in Buffalo Reproduction and AI.FAO Animal Production and Health Paper 13. pp. 173–181. FAO (Food and Agriculture Organization of the United Nations), Rome, Italy.

Cockrill WR. (1974). The husbandry and health of the domestic buffalo. Food and Agriculture Organization of the United Nations, Rome.

Corbet GB and Hill JE. (1987). A World List of Mammalian Species, Second edition. London: British Natural History Museum. ISBN 0565009885.

Deb RN and Kadu MS. (1977). Some observations on production characters in Nagpuri buffalo. Indian Journal of Animal Health 16:35–38.

FAO (Food and Agriculture Organization of the United Nations), FAOSTAT Agriculture Data, 2004. http:\\apps.fao.org/default.htm.

Gentry A, Clutton-Brock J and Groves CP. (2004). The naming of wild animal species and their domestic derivatives. Journal of Archaeological Science 31: 645–651.

Groves CP. (1971). "Request for a declaration modifying Article 1 so as to exclude names proposed for domestic animals from Zoological Nomenclature." Bulletin of Zoological Nomenclature 27: 269–272.

Hafez ESE. (1955). Puberty in the buffalo cow. Journal of Agricultural Sciences 46: 137–142.

Khan MA, Islam MN, Siddiki MS. (2007). Physical and chemical composition of swamp and water buffalo milk: a comparative study. Italian Journal of Animal Science 6: 1067–1070.

Chapter 2

Ovarian Development and Oogenesis

Introduction

Ovaries are a pair of organs found in lumbar region of abdominal cavity at a short distance caudal to kidneys. They are covered by peritoneum and are suspended by mesovarium from the body wall. The mammalian ovary serves as the female gonad responsible for development and release of a mature oocyte for fertilization and successful propagation of a species. It also serves as an endocrine organ producing steroid hormones that lead to development of female secondary sexual characteristics and establishment of estrous cycle and pregnancy. Buffalo ovary has an oval shape and is smaller than cow ovary. The mean length, width and diameter of adult buffalo ovary measures 2.46 ± 0.07, 1.77 ± 0.07 and 1.50 ± 0.04 cm, respectively, while the mean weight ranges from 3 to 5 g. The outermost layer of an ovary consists of germinal epithelium, directly underneath of which is a layer of dense connective tissue known as tunica albugenia. Under the tunica is located ovarian cortex, a conjunction of ovarian follicles, surrounding fibroblasts, collagen and elastic fibers. The medulla is the innermost portion composed of blood vessels, lymphatic vessels and nervous terminals.

Development of buffalo ovary

The formation of functional ovary depends on three major physiological events during early stages of oogenesis: i) Initiation of meiosis; ii) Formation of follicles; 3) Development of hormone producing cells. It could broadly be classified into pre-natal and post-natal/pubertal stages of development.

Pre-natal development

Cells that finally give rise to oocytes are known as primordial germ cells (PGCs) and their lineage constitutes the germ line. These cells are differentiated from a subset of epiblast (inner cell mass) cells of an implanted embryo and can be identified in the yolk sac at 4th week of gestation. The cells are induced by bone morphogenetic protein (BMP4, BMP8, BMP2) signaling from the extra-embryonic ectoderm. BMP signaling induces *FRAGILIS* and *BLIMP1* gene expression which cause inhibition of somatic differentiation and preservation of germ cell pluripotency, during a developmental stage at which strong morphogen gradients act to pattern the early embryo. Following their specification, PGCs under the influence of *STELLA*, *c-KIT* and *OCT4* proliferate and migrate towards the sexually undifferentiated gonad to colonize it and partner with its developing somatic supporter cells. The migration takes place at 4–6 weeks of gestation during which PGCs move from yolk sac to gut tube wall and from there to dorsal body wall. Here they come to rest on either side of midline in the loose mesenchymal lining of the coelomic cavity. This stimulates the adjacent coelomic epithelium to proliferate and form somatic support cells. Proliferation of the somatic support cells create a swelling just medial to each mesonephros on both left and right sides of gut mesentry. This swelling is known as *genital ridge* and represents the primitive/ indifferent or bipotent gonad. The somatic support cells invest PGCs and form tissues that nourish and regulate development of maturing sex cells - ovarian follicles. The relationship between somatic cells and PGCs is bi-directional implying that somatic cells are essential for germ cell development within the gonad but if PGCs fail to arrive in the presumptive gonadal region, the gonadal development is disrupted and somatic cells die.

A number of hypothesis have been proposed for PGC migration like: a) Active migration by developing pseudopodia; b) Migration under influence of chemotactic signals by chemotrophic agents or chemokines like stromal cell-derived factor-1a (SDF-1a) secreted by surrounding somatic cells; c) Migration by substrate-guided mechanism where PGCs migrate in direction of stress fibers in underlying somatic cells. This pathway is thought to be mediated by proteoglycans like Syndecan-4, Fibronectin, Hyaluronan, Versican, Tenascin-C and other extra cellular matrix molecules like Collagen type I and III. Irrespective of the mechanism, the migration process is divided into three distinct phases: i) *Separation phase* - when PGCs leave the hindgut epithelium and enter the mesenchyme of the dorsal mesentry; ii) *Migration phase* - PGCs move between mesenchymal cells of the dorsal mesentry and travel towards the genital ridge; iii) *Colonization phase* - when PGCs reach and populate the genital ridges.

The PGCs by now are called as *gonocytes*. They undergo a few more mitotic divisions after which they are invaded by somatic support cells. By the third month of gestation, the gonads are at a late stage of differentiation. The primitive germinal, granulosa, thecal and stromal cells are present but tunica albugenia is poorly developed. The germinal cells are mostly oogonia, majority of them still in mitosis, though a few ocytes could be seen in early meiosis. By 4 months of gestation, the germinal epithelium invades the gonad to form epithelial cords and reteovaries are apparent. The germinal cells are still at the oogonial stage and many, particularly those in the center of the organ, degenerate. This represents the *first wave* of germinal cell degeneration. The ovarian stroma and tunica albugenia become more organized by the 5th month when sex cords invade the stroma. The meiotic activity becomes apparent only after formation of primordial follicles begin, suggesting an important role of granulosa cells in initiating meiosis. The degeneration of oogonia and the primordial follicle formation continues up to 6th month of gestation. The primordial follicle consists of a dictyate-oocyte surrounded by a single layer of flattened pre-granulosa cells. This degeneration represents the *second wave* of germinal cell degeneration. The primordial germ cells enclosed by granulosa cells are found to survive more likely than the naked germinal cells, indicating the necessity of folliculogenesis. Oocytes and granulosa cells are both interdependent and interconnected, through connexins 37 and 43 and oocyte requires these cells to grow and survive. It is presumed that primordial follicle formation stimulates ovarian growth. The organization of follicles characteristically occurs predominantly at the cortex of ovary than in the center where more degeneration occurs. This gives rise to "whirl" like structures which arise primarily due to stromal cell abundance. By 5–6th month of gestation, primordial follicles are activated into primary follicles, characterized by acquisition of a complete layer of 11–20 cuboidal granulosa cells around oocytes and the activation of primordial follicle growth. During 7 and 8 month of fetal age, tunica albugenia reaches its maximum thickness (200 µm) and the cells of germinal epithelium become flattened. Most of the germinal cells are oocytes now and several oocytes undergo degeneration to form "*Z cells*." The germinal cells attain their maximum volume by 8th month of gestation. The peak degeneration of oocytes at pachytene also occurs at this time, representing the *third wave* of degeneration. By 9th and 10th month of gestation, vesicular and atretic follicles form, primarily due to increase in maternal plasma of follicle stimulating hormone (FSH) levels. The ovary is more clearly divided into cortical and medullary zones with germinal cells only in the former. There is an abrupt increase in ovarian weight from 8th to 10th month of fetal age. Thus, the pre-natal ovary contains four distinct cell types: i) Primordial germinal cells; ii) Granulosa cells; iii) Thecal cells; iv) Stromal cells.

Post-natal development

In fetal ovary all the oogonia have entered into meiosis to form primary oocytes. The meiosis, however, is arrested near the end of gestation period and the primary oocytes enter a state of dormancy for a prolonged period of time. The arrest is overcome at sexual maturity in response to oestrogen, FSH and LH (luteinizing hormone). The cyclical activity of these hormones sets in menstrual cycle and folliculogenesis in various follicles, one among which completes the first meiotic division to finally undergo ovulation. During oocyte maturation, the interval between LH surge and ovulation, oocyte completes meiosis I, forms extensive cellular interactions with granulosa and theca cells, undergoes asymmetric cytokinesis and extrudes a smaller haploid polar body. All these processes occur primarily in one follicle, known as the dominant follicle, which is selectively recruited for growth. Once a follicle has become recruited for growth, however, there is a burst of transcription and translation in its oocyte, and this is sustained until the cell reaches full size. The meiosis is now arrested in metaphase II and this stage is known as Germinal vesicle stage. Meiosis is resumed after fertilization by the sperm. This completes the release of second polar body and formation of a ruptured follicle which either undergoes atresia or develops into corpus luteum, depending on whether the pregnancy is established or not.

Follicular dynamics

Folliculogenesis implies to progression of a number of small primordial follicles to large preovulatory follicles that enter the menstrual cycle. It concerns to a lengthy developmental process a follicle goes through, from the time it leaves the reserve pool and begins to grow by cell proliferation and antrum formation until ovulation or atresia. Folliculogenesis starts before birth in some mammalian species (cow, sheep and buffalo) or shortly after birth in other species (mouse, rat and hamster). In a larger perspective, the whole folliculogenesis, from primordial to preovulatory follicle, belongs to the stage of *otidogenesis* of oogenesis.

After the primordial germ cells invade genital ridge and initiate its growth and development, they themselves undergo a series of mitotic divisions and get invested by somatic support cells to form oogonia. The first oogonia that undergo meiotic division are located in the innermost areas of ovarian cortex from where developmental waves of meiosis spread outwards. By mid-to-late-gestation many stages of germ cell development are simultaneously present in the fetal ovary. The oogonia enter the first meiotic division by 3–4 weeks of gestation and then immediately become dormant, by ceasing the meiosis at prophase of meiosis I. The oocyte at this stage is known as primary oocyte. The nucleus of these dormant primary oocytes contains partially condensed prophase chromosomes. It becomes very large and watery and is referred to as germinal vesicle. The swollen condition of the germinal vesicle is attributed to protect the oocyte DNA during the long period of mitotic arrest. The oocyte remains as primary oocyte, in dormant state, until puberty, when at each cycle selected follicle(s) go on to ovulate. Some of the primary oocytes die by apoptosis before birth and females are always born with a lesser number of oocytes than the maximum number reached during fetal life. For example, buffalo ovary has maximum number (~ 23,540) of germ cells at 7 months of gestation which reduces to 20,000 at birth while the cattle ovaries contain 2,7,00,000 at 3–4 months of fetal life which reduces to 68,000 at 13 days after birth. A single layer of squamous epithelial cells tightly encloses each primary oocyte forming the primordial follicle. Thus primordial follicle is characterized by a quiescent oocyte, arrested in prophase I of meiosis and surrounded by a single layer of flattened granulosa cells. These primordial follicles constitute the ovarian reserve and get engaged for further development of the ovary. The most abundant organelles found in primordial follicle oocytes are round-shaped mitochondria which are known to be immature forms of this organelle. They develop into elongated shape as they mature. The presence of immature mitochondria is consistent with primordial follicle containing a quiescent oocyte that does not require a large amount of energy to survive. The primary oocyte ooplasm also contains lipid droplets, endoplasmic reticulum, golgi cisternae, polyribosomes etc. Buffalo ooplasm at this stage has a characteristic region containing well-developed smooth endoplasmic reticulum. In primordial follicles, granulosa cells are small and have a relatively large nucleus that matches the cell format. In goat granulosa cells present low density of cytoplasmic organelles, and in buffalo scarce myelin figures are present which are presumed to be the result of the digestion of old or nonfunctional structures. There are no specialized junctions between granulosa cells or between them and the oocyte. The exchange of substances between the cell types occurs by endocytosis or by diffusion. Initiation of growth and transition from the primordial to primary follicle begins with the *recruitment* of primordial follicles. At this stage the follicles become, "committed" and growth proceeds until either up to ovulation or atresia. The growth takes place in only a small number of follicles at each time and the factors responsible for triggering the development remain one of the major unsolved problems of ovarian physiology. Primordial follicle leads to formation of primary follicle which is characterized by differentiation and proliferation of granulosa cells and enlargement of the oocyte diameter from 35 to 42 μm. Granulosa cells increase in number (from 4–8 in primordial follicle to 8–20 in primary follicle), become cuboidal in shape and form adherens junctions between themselves and with the oocyte. The oocyte nucleus becomes irRegular and ooplasm contains round mitochondria, endoplasmic reticulum, golgi cisternae and vesicles. The oocyte undergoes volume expansion and zona pellucida proteins start to be secreting between the growing oocyte and granulosa cells. The mitochondria start dividing and are a bit elongated. The primary follicle starts transition to secondary follicle which is characterized by further growth (upto 53 μm diameter) and oocyte volume. The organelles migrate to the periphery of ooplasm, leaving an organelle-free zone next to nucleus. The oocyte is predominantly spherical with the ooplasm containing vesicles, round and elongated mitochondria in cow, sheep, goat, cat, buffalo, human and yak. The most abundant organelle is mitochondria, with more frequent elongated forms implying to higher metabolic activity of the oocyte at this stage. The number of cytoplasmic vesicles increases in active oocytes, occupying most of the ooplasm, denoting the stock of different biomolecules like proteins, polysaccharides or even lipids. Zona pellucida is usually formed around the oocyte in secondary follicles, though it is visible in only patches in the buffalo secondary follicles, indicating incomplete formation. Gap junctions are formed between oocyte and granulosa cells and these are considered to be essential for oocyte growth and metabolism. Cortical granules are seen for the first time in secondary follicles. They are vesicular structures containing enzymes that undergo exocytosis upon fertilization. The cortical granules are aligned near the oocyte plasma membrane and the release of their contents aims to harden the zona pellucida to prevent polyspermy. The cortical granules usually appear in clusters, either distributed throughout the ooplasm or confined to deep cortical area near the Golgi complex. Granulosa cells at this stage have been implicated to be involved in steroidogenesis and synthesis of zona pellucida. Stroma-like theca cells are recruited by oocyte-secreted signals, which surround the follicle's outermost layer, the basal lamina. The theca layer is still poorly developed in small secondary follicles but in large secondary follicles a clear theca interna layer is present. The space between adjacent granulosa cells begins to be filled with follicular fluid, whose progressive accumulation causes distension of the cavities and initial formation of antrum. This leads to formation of antral follicles. These follicles are also known as Graafian follicles. This transition from preantral to early antral follicle is a critical stage of follicular development in terms of follicle destiny, *viz.*, growth or atresia. The interaction between granulosa (oocyte cells) and somatic cells (theca cells) is especially important for further follicular development. Antrum formation occurs in cattle and buffalo follicles when they reach 120–160 μm diameter in contrast to pig and sheep follicles, which undergo antrum formation at 400 and 220 μm diameter, respectively. The difference in the timing of antrum formation is important in the overall process of folliculogenesis, since there is a substantial increase in growth of follicles after antrum formation. The fluid-filled antrum now separates the cumulus-oophorous cells surrounding the oocyte from granulosa cells lining the wall of the follicle. The secondary follicle attains a tremendous size and becomes dependent on hormones like Follicle-stimulating hormone. Granulosa and theca cells continue to undergo mitosis and the follicle now becomes tertiary follicle. The oocyte is completely surrounded by zona pellucida, which is crossed by granulosa cell projections forming indentations in the oolema. The organelles are evenly distributed in ooplasm, mitochondria are elongated, and lipid droplets and vesicles increase in number. In buffalo oocytes at tertiary follicle stage, organelles are located in the perinuclear region, mitochondria in the cortical area and lipid droplets in medullary area. This organization is considered as an indicator of high metabolic rate, which tends to increase with development and growth. Under the action of oocyte-secreted morphogenic gradient, the granulosa cells of the tertiary follicle undergo differentiation into four distinct subtypes: *corona radiata* cells that surround zona pellucida; *membrana* cells that are located interior to basal lamina; *periantral* cells that are adjacent to antrum; and *cumulus oophorous* cells which connect the membrana and corona radiata granulosa cells together. Follicular development till this stage is considered to be gonadotrophin-independent and is essentially driven by locally secreted factors. Further oocyte development or death is gonadotrophin-dependent as follicles now express functional follicle stimulating hormone (FSH) and luteinizing hormone (LH) receptors. Each type of cell behaves differently in response to FSH. For example, in response to rise of FSH antral follicles begin to secrete estrogen and inhibin, which in turn exert a negative feedback on FSH secretion. Theca cells express receptors for luteinizing hormone (LH) and under its action produce androgens, like androstendione, which are aromatized by granulosa cells to produce estrogens, primarily estradiol whose level begins to rise during subsequent growth. Follicles with fewer FSH-receptors are unable to develop further, thereby showing retardation in their growth rate and finally undergo atresia. Eventually only one follicle remains viable. This follicle is known as dominant follicle. It grows quickly and achieves tremendous size to become preovulatory follicle. The LH surge induces what we call as oocyte maturation in the dominant follicle destined to become ovulatory follicle. In ovulatory follicle cortical granules move towards subplasmalemal area and migrate towards periphery of the oocyte. The oocyte gets activated and completes meiosis I, again to get arrested in metaphase II phase of meiosis II. At the end of maturation period, when oocyte reaches metaphase II, cortical granules are aligned to the inner surface of the oocyte plasma membrane ready to release their contents as soon as the oocyte is fertilized, to prevent polyspermy. The maturation is achieved when oocyte reaches approximately 80% of its final size. Association between endoplasmic reticulum, mitochondria and lipid droplets becomes common. The association is related to lipid metabolism and endoplasmic reticulum-mitochondria calcium signaling. It allows efficient transmission of signals from cytosolic calcium to mitochondria, enabling activation of mitochondrial metabolism and an increase in ATP supply from the calcium pump in endoplasmic reticulum. It is likely that in oocytes at this stage of development, this structure is involved in regulation of sperm-triggered calcium oscillations. The membranes of the endoplasmic reticulum are physiologically active and interact with cytoskeleton. Cumulus oophorus layer of the preovulatory follicle develops an opening or stigma to excrete the oocyte with a complement of cumulus cells. This process is called as ovulation. The follicular phase of menstrual cycle represents the time between selection of a tertiary follicle and its subsequent growth into a preovulatory follicle. The mature oocyte is finally ovulated, having extruded the first polar body. Thus folliculogenesis constitutes morphological, biochemical and molecular modifications, which lead to nuclear and cytoplasmic maturation of an oocyte and guarantee its competence for fertilization. The ruptured follicle undergoes a dramatic transformation into corpus luteum (CL). CL functions as a transient endocrine gland and plays a critical role in regulation of various reproductive processes like control of estrous cycle, embryonic development, implantation and maintenance of pregnancy. CL is composed mainly of *steroidogenic* and *non-steroidogenic* cells and is mainly responsible for production of progesterone. Steroidogenic cells include large and small luteal cells whereas endothelial cells, fibroblasts, macrophages and other cell types have been grouped together as non-steroidogenic. The small and large luteal cells originate from theca and granulosa cells, respectively. With the advancement of pregnancy, transformation of small luteal cells to large luteal cells occurs. The functioning of CL depends on interactions among its cellular components in an autocrine / paracrine fashion which produces different steroidogenic and non-steroidogenic factors. The development and size of CL also depends upon the seasonal variations; for example, buffalo CL size is smaller (0.94 cm) during summer than in winter (1.27 cm). The size as well as weight of buffalo CL is lesser than cattle CL which may contribute to greater rate of embryonic mortality. CL of buffalo is also embedded in buffalo ovary in contrast to cattle CL which are more superficially located.

Follicular waves

The female gonad has been a matter of much controversy throughout its study from 5th century B.C. when Hippocrates did not ascribe any generative role to the organ. He rather suggested that new life results from the action of two kinds of semen- one from the male (ejaculate) and other from the female (menstrual blood). A century later Aristotle thought of the ovary as an imperfect vestige of the male testis with no apparent function. The concept was held up to mid 1600s when ovary was recognized for what it was - the producer of eggs and an equivalent female organ to male testes. The Dutch physician, Regnier de Graaf, is often cited as the first to recognize the rightful role of the ovary in his "New treatise concerning the generative organs of women" which was published in 1672. The common belief was held that follicle itself was the egg like a small bird's egg without a shell until Karl Ernst in 1827 provided the first description of a mammalian egg from his microscopic study of ovarian follicles in dog ovary. The first studies on the dynamics of follicle development, however, were not for another 100 years. By now we know that oocytes originate from primordial germ cells from the endoderm of the embryonic yolk sac, migrate by amoeboid movement *via* the dorsal mesentry of hindgut to colonize into the gonadal ridge. These PGCs are internalized into the gonadal ridge through its surface epithelium which was initially thought to be the source of PGCs and mistakenly named the germinal epithelium. PGCs cease mitotic divisions during the process of internalization and become internalized in germ cell cords (ovigerous cords) composed of epithelial cells which are delineated from the surrounding mesenchymal cells by a basal lamina. The PGCs are now referred to as oogonia. The oogonia start meiosis and are now referred to as primary oocytes when the meiosis is halted at pachytene stage of prophase-I. The chromosomes are decondensed and contained in the nuclear membrane (Germinal vesicle). This is followed by enclosure of primary oocyte by a single layer of flattened epithelial cells (from germ cell cords) to form primordial follicles. Those oocytes which fail to be surrounded by epithelial cells degenerate. The initiation of follicular growth (activation) begins with the transformation of the flattened pre-granulosa cells of the primordial follicle into cuboidal granulosa (follicular) cells of the primary follicle. The granulosa cells undergo proliferation and form 2 to 6 layers around the oocyte (now known as secondary follicle). This is followed by formation of tertiary follicle in which the oocyte is surrounded by more than 6 granulosa layers. The follicle thereafter develops antrum which is filled with follicular fluid, forming a structure known as antral follicle. The development of a primordial follicle and its progression to pre-ovulatory and finally to ovulatory follicle (follicle capable of releasing the oocyte for fertilization) is simply referred to as folliculogenesis. It is a highly controlled and dynamic process which begins continuously from sexual maturity till menopause. It stars from a primordial follicle, selected randomly from the fixed reservoir pool (10,000 to 20,000 in buffalo) and ends with the formation of a preovulatory follicle capable of releasing an oocyte in a hormonally controlled process of ovulation. The process of folliculogenesis coincides with the corresponding estrous cycle. Endocrinological and ultra-sound studies have demonstrated that follicle growth occurs in waves during estrous cycle of bovines and buffaloes. In each wave of follicular growth, one dominant follicle develops and suppresses the other follicles. The dominant follicles grow and reach maximum diameter in the middle of the estrous cycle. The dominant follicle that develops during the last wave of follicular growth in each estrous cycle becomes the ovulatory follicle.

Follicular wave pattern in buffalo

A wave of follicular development consists of simultaneous growth of a large number of small antral follicles (follicular recruitment), selection of a dominant follicle and regression of subordinate follicles. This phenomenon is regulated by a follicular selection mechanism which causes atresia of all the follicular cohorts. When there is no luteal regression, the dominant follicle becomes atretic and a new follicular wave begins. The determination of the pattern of follicular waves during the estrous cycle and number of recruited follicles per wave are important for improving the response to super ovulation treatment. Since, the wave pattern does not seem to be related to a breed but is more of an individual property; the assessment of follicular wave pattern in an individual buffalo, by monitoring the follicular development with an ultrasound machine, assumes greater significance. A marked individual variation exists in follicular dynamics among buffaloes, with as few as one to as many as three waves of follicular growth occurring within an estrous cycle. A group of reproductively sound buffaloes, showing similar wave-pattern should then be subjected to super ovulation treatment for better results. The selection of animals based on number of follicles recruited per wave is encouraging because of the positive correlation found between the number of small follicles at the beginning of super-ovulatory treatment and response. The treatments for estrous synchronization and super ovulation might also be optimized to the number of follicular waves that occur in an oestrous cycle.

It has been shown that buffalo have estrous cycles with 1, 2 or 3 follicular waves. The 2 wave cycles are the most common. The number of waves in a cycle is associated with the luteal phase and estrous cycle length. While defining a follicular wave, the day of ovulation is usually defined as Day 0. The first wave commences on day 1 in all categories of animals, while the second wave emerges usually on days 10.8 and 9.3 for the animals with 2 and 3 wave cycles, respectively. The third wave usually emerges on day 16.8 in 3 wave estrous cycle. The estrous cycles with two and three waves of follicular development differ in mean length of the luteal phase (10.4 compared to 12.7 days) and the interval between ovulations (22.3 compared to 24.5 days). For buffalo with two-waves of ovarian follicular development during estrous cycles, the growth rate and diameter of the largest follicle is smaller in heifers as compared to cows for both the first and second follicular wave. There has been no report of 4 follicular waves within an estrous cycle in buffalo, although 4 waves do occur in cattle.

A follicular wave begins (emergence) when a number of small follicles are recruited for growth (recruitment) at day 1 after ovulation. A group of antral follicles is formed which attain a diameter of 4 mm. At this stage, one of the follicles, termed as dominant follicle, continues to grow while others become subordinate follicles and initiate a process of atresia. The transformation of primordial follicles through the growing stage to the tertiary stage is very inefficient in buffalo and could be viewed as a response to estrogen and progesterone levels within the follicle. The molar ratio of estrogen to progesterone is an indicator of a follicle's fate, growth or atresia. The buffalo ovarian follicles could be classified as estrogen active (estrogen to progesterone molar ratio >100) and estrogen inactive/ atretic (estrogen to progesterone molar ratio < 100). Selection of the dominant follicle is, therefore, associated with a deviation in growth rate between the dominant and the largest subordinate follicle, an event termed *deviation*. The maximum size of each dominant follicle is around 10–15 mm and exceeds the diameter of all other follicles. Subordinate follicles (follicular cohorts) are those follicles that appear from the same follicular pool as the dominant follicle but lose the race to the fastest growing dominant follicle. The dominant follicle and its cohort of subordinate follicles is defined as a wave. The first wave usually recruits more follicles for growth than the second or third wave. 1 follicular wave cycle produces only one dominant follicle while 2 and 3 wave patterns produce 2 and 3 dominant follicles, respectively. It has been reported that the first dominant follicle persists for a longer time in 2-wave than in 3-wave cycles. The maximum diameter of first dominant follicle is not different from the ovulatory follicle in 2-wave cycle, while in 3 wave cycle the first dominant follicle is larger than the second dominant follicle. The onset of second and third waves in 2 and 3 wave cycles begins 2 or 3 days after the beginning of the static phase of the previous dominant follicle. Thus, static phase ends the dominance of the dominant follicle. The 2 and 3-wave cycles differ significantly with regard to: i) persistence of the first dominant follicle (20.7d *vs* 17.9 d); ii) length of growth phase (7.39d *vs* 5.50 d); iii) static phase duration (6.88 d *vs* 5.30 d); iv) last day of growth phase (8.55 d *vs* 6.60 d); v) beginning of regression phase (15.4d *vs* 11.9 d); vi) maximum diameter of the first dominant follicle (1.51 cm *vs* 1.33 cm); vii) maximum diameter of the ovulatory follicle (1.55 cm *vs* 1.34 cm).

It has been observed that the number of follicles recruited in a follicular wave is lower in buffaloes as compared to cattle. This could be due to small pool of primordial follicles in buffalo ovary as compared to cattle. Buffalo also exhibit a greater rate of follicular atresia (67%) than cattle (50%). Buffalo ovary possesses lesser number of antral follicles as compared to the cattle ovary (20% of cattle).

Further reading

Barkawi A, Hafez Y, Ibrahim S, Ashour G, El-Asheeri A, Ghanem N. (2009). Characteristics of ovarian follicular dynamics throughout the estrous cycle of Egyptian buffaloes. Animal Reproduction Science 110: 326–334.

Baruselli P, Mucciolo R, Visintin J, Viana W, Arruda R, Madureira E, Oliveiras C and Molero-Filhof J. (1997). Ovarian follicular dynamics during the estrous cycle in buffalo (bubalus bubalis). Theriogenology 47: 1531–1547

Fernanda P, Renata C, José L and Carolina M. (2014). Ultrastructural changes in oocytes during folliculogenesis in domestic mammals. Journal of Ovarian Research 7:102 http://www.ovarianresearch.com/content/7/1/102.

Flor Sánchez and Johan Smitz. (2012). Molecular control of oogenesis. Biochimica et Biophysica Acta 1822: 1896–1912.

Giuseppina M. (2012). Ultrasonography and reproduction in buffalo. Journal of Buffalo Science 1: 163–173.

Ian M, Christiane R I, Lindsay U G, Cássia M,Eunice O. (2008). Ovarian follicular dynamic during early pregnancy in buffalo Bubalus bubalis heifers. Ciência Animal Brasileira 9: 121–127.

Perera B. (2011). Reproductive cycles of buffalo. Animal Reproduction Science 124: 194–199.

Syed Mohmad Shah, et al. (2015). Comparative expression analysis of gametogenesis associated genes in bubaline (B. Bubalis) ovaries and testes. Reproduction in Domestic Animals 50: 365–377.

Taneja M, Ali A and G. Singh. (1996). Ovarian follicular dynamics in water buffalo. Theriogenology 46:121–130.

Chapter 3
Buffalo Oestrous Cycle

Introduction

Estrous (oestrous) cycle comprises the recurring physiological changes that are induced by reproductive hormones in most mammalian therian females. The cycle consists of repetitive and predictable reproductive events which start after sexual maturity in females and continue till death, with interruptions of anestrus or pregnancies or seasons of the year. The cyclicity may also cease if nutrition is inadequate or environmental conditions are unusually stressful or in certain pathological conditions. The term "estrous" is derived from Latin word *oestrus* meaning *frenzy* or *gadfly*. It has been used to describe *rut in animals* or *heat.* While most placental female mammals experience an estrous cycle, humans and great apes (chimpanzees, gorillas and orangutans) experience a true menstrual cycle. Estrous cycle differs from menstrual cycle in that: i) Animals with estrous cycle reabsorb the endometrium if conception does not occur during that cycle, while animals with menstrual cycle shed the endometrium through menstruation instead; ii) In estrous cycle females are generally only sexually active (in heat) during the estrus phase of their cycle, while females with menstrual cycle can be sexually active at any time in their cycle, even when they are not about to ovulate. This leads to formation of pair bonds between a male and a female of the species; iii) Menstrual cycle is not characterized by such phases of heat during which the female shows obvious external signs for mating.

Frequency and timing of estrous

The frequency of estrous cycle varies with species. Some species experience Regular uninterrupted cycles throughout the year or for a specific period of the year. Such species like rodents, cat, cattle and swine are said to be polyestrous. These animals have uniform distribution of estrous cycles that occur Regularly throughout the year. They can go into heat several times a year, depending on the duration of their estrous cycle. The polyestrous females can become pregnant throughout the year without regard to season. Some animals are seasonally polyestrous or seasonal breeders meaning that they exhibit more than one estrous cycle during a specific time of the year. These animals display clusters of estrous cycles that occur only during a certain season of the year. These are usually divided into: a) Short-day breeders like sheep, goat, deer etc. which are sexually active in fall or winter; b) Long-day breeders like horses, hamsters and ferrets which are sexually active in spring and summer. Monoestrous species, like bear, fox, wolf, Great panda, etc. display only one breeding season in a year. The breeding season typically falls in spring so as to allow growth of the offspring during the warm season to aid survival during the next winter. There are also some Diestrous species like domestic dog which go into heat only twice per year. Some mammalian species like rabbit do not have an estrous cycle and are able to conceive at any arbitrary moment. Rabbit is also classified as an induced or reflex ovulator in that it requires mating in order to ovulate in contrast to spontaneous ovulation in farm animals. In spontaneous ovulation, follicle rupture follows a series of hormonal events which induce a pattern of GnRH release that stimulates a pre-ovulatory LH surge, while in induced ovulation coitus provides a discharge of GnRH that causes LH release and ovulation, usually 12 h after mating. A few species undergo behavioral estrous and ovulation in response to certain stimuli. The females require physical contact to begin estrus and mating to promote ovulation. For example certain species of voles, like prairie voles (*Microtus ochrogaster*) and montane voles (*M.montanus*), remain reproductively inactive until there is a stimulus from a male.

The estrous cycle allows animals to place most of their mating energy into particular periods and provide obvious signs to the male partners for successful mating. The repetitive nature of estrous cycle together with the overt external signs of heat is one of the most watched for natural mating. The timing of estrous cycle is generally coordinated with seasonal availability of food, shelter, weather and other circumstances like migration, predation, etc. Each aspect is considered so as to maximize the offspring's chances of survival. Some species are able to modify their estral timing in response to external conditions.

Reproductive anatomy of buffalo

The reproductive tract of the water buffalo is quite similar to that of domestic cattle. However, the tubular genitalia of buffalo are more muscular and firmer with more coiled uterine horns than those of the cow. The uterine body is much shorter (1–2 cm) than that of cow (2–4 cm), cervix is smaller (3–10 cm length and 1.5–6 cm diameter) with more tortuous canal which accounts for less dilation of the external os during estrus. Buffalo, on an average has 3 cervical folds in contrast to 4–5 in cow. Buffalo also have smaller ovaries than cattle. The inactive ovaries of a mature water buffalo measure 3 cm x 1.4 cm x 1 cm versus 3.7 cm 2.5 cm x 1.5 cm in case of cow. Buffalo ovary weighs 2.9 to 6.1 g while that of cow weigh 5 to 15 g. The ovaries of post-pubertal buffalo heifers have a reservoir of only 10,000 to 20,000 primordial follicles compared to over 100,000 in cattle. The mature ovaries weigh around 2.5 g when inactive and 4 g when active. When palpated per rectum, mature follicles rarely exceed 8 mm in diameter and tend to protrude from the surface of the ovary. This corpus luteum is smaller than that in cattle and often does not protrude markedly from the surface of the ovary and sometimes lacks a clear crown. These characters make accurate identification of ovarian structures difficult. The mature follicles measure 1.3 to 1.6 cm in diameter, while mature corpus lutea measure 1.2 to 1.7 cm in diameter versus 1.2 to 2.5 cm in cattle, as analyzed by ultrasonic imaging.

Factors affecting puberty

Buffalo heifers usually attain puberty when they reach about 55–60% of their adult body weight but the age can be highly variable ranging from 18 to 46 months. A number of factors like genotype, nutrition, management, social environment, climate, year and season of birth and disease determine the age of puberty. Under favourable conditions riverine buffalo exhibit first estrous at 15–18 months of age, while swamp type does so at 21–24 months. The body weight at which puberty is attained is strongly influenced by genotype and is around 200–300 kg for swamp type and 250–400 kg for the river type. Although buffalo attain puberty later than cattle (6–12 months of age) they have a longer reproductive life which tends to compensate for this early economic disadvantage.

Reproductive patterns in buffalo

Buffalo are polyestrous and are capable of breeding throughout the year. However, a seasonal pattern of breeding activity and calving has been observed in many countries. In tropical countries with relatively constant photoperiod annual changes in rainfall appear to influence estrous cyclicity, with availability and quality of herbage related to this cyclical reproductive pattern. In dry zones, buffalo commence ovarian activity some 2–3 months after the onset of monsoonal rains. In temperate regions where buffalo are fed with constant balanced diet; photoperiod, and not the diet, is the main reason of seasonal reproductive pattern which is mediated by melatonin secretion. Heat stress during the hot summer months in India is a major cause of anestrous in buffalo. The stress elevates blood concentrations of prolactin which as a result of decreasing progesterone secretion influences ovarian activity causing sub-fertility and repeat breeding.

Estrous cycle

Walter Heape originally described estrous cycle as a continuous sequence of five events viz. proestrus, estrus, metestrus, diestrus and anestrus. The terminology is helpful for laboratory rodents for which it was proposed but it has been found not to be helpful for describing events in larger domestic animals. The events, nevertheless, require a little description as most of the estrous cycle descriptions still hold the terminology, even in domestic animals.

Proestrus: This refers to first phase of estrous cycle during which one or several primordial follicles, depending on the species, start growing. The phase can last from as little as one day to as long as 3 weeks. During this phase, endometrium starts to develop and animals may experience vaginal secretions that could be bloody. The female is not sexually receptive yet.

Estrus: Under the influence of gonadotropic hormones, especially estrogen, ovarian follicle continues growing. The animal exhibits sexually receptive behavior and is referred to be in heat. Its duration varies from 1–7 days depending upon the species, while in induced ovulators, if no copulation occurs, it may continue for many days, followed by *interestrus* phase. The estrus phase starts again until copulation and ovulation occurs.

Metestrus: The signs of estrus subside during this phase and corpus luteum starts to form. The endometrium begins to secrete small amounts of progesterone. The phase can last for 1 to 5 days and may be accompanied by bleeding due to declining estrogen levels.

Diestrus: This phase is characterized by activity of corpus luteum that produces progesterone. In absence of fertilization, the phase (also termed pseudo-pregnancy) terminates with regression of corpus luteum. The lining of the uterus is not shed but is reorganized for the next cycle.

Anestrus: This refers to a phase when sexual cycle rests. This is typically a seasonal event controlled by many factors like photoperiod, pregnancy, lactation, illness and age.

In buffalo the duration of estrous cycle is similar to that in cattle, ranging from 18 to 24 days, with a mean of around 21 days. A great variability of estrous cycle length has however, been observed in buffalo with a greater incidence of both abnormally short and long estrous cycles. This is attributed to various factors like adverse environmental conditions, nutrition and irRegularities in secretion of ovarian steroid hormones. The cycle has been divided into two discrete phases: i) *Luteal phase* – its duration is 14–18 days after ovulation and is associated with formation of corpus luteum. It is often divided into met-estrus and di-estrous; ii) *Follicular phase* - this follows after demise of corpus luteum until ovulation. It is further divided into proestrus and estrus. This phase is associated with final maturation of ovulatory follicle and ovulation. The duration of estrus is similar in river and swamp buffalo, varying between 5 and 27 h. Ovulation occurs about 24–48 h (average 34 h) after onset of estrus or 6 to 21 h (mean 14 h) after end of estrus. In hot climates, duration of estrus tends to be shorter and signs of estrus may be exhibited only during night or early morning. In Italian buffalo different durations of estrus have been observed and categorized as short (< 12 h), medium (13–24 h), long (24–48 h) and very long (> 48 h). In short and medium estrous cycles ovulation occurs after the end of estrus, around 6–72 h and 24–60 h, respectively, while in long and very long estrous cycles ovulation occurs before the end of estrus.

Physiology and endocrinology of estrous cycle

Based on the physiological and endocrinological events, buffalo estrous cycle is also divided into 4 phases: **estrus** (day 0), **metestrus** (day 1--4), **di-estrus** (day 5 – 18) and **pro-estrus** (day 19 to ovulation). This is further simplified into **follicular stage** representing the time of greatest follicular growth, including behavioural estrus and ovulation; and **luteal stage** characterized by maximum growth and maintenance of corpus luteum and progesterone production. The different phases of reproductive cycle are regulated by intricate sequential events and interaction between hypothalamic releasing factors from the pituitary and sex steroids from ovary and endometrium. The overall regulation of estrous cycle is due to integration and synchronization of various endocrinological factors like gonadotrophin releasing hormone (GnRH) from hypothalamus, follicle-stimulating hormone (FSH) and luteinizing hormone (LH) from anterior pituitary, progesterone (P4), estradiol (E2) and inhibin from ovaries, and prostaglandin (PGF2α) from uterus. These hormones function through a system of positive and negative feedback to govern the estrous cycle. There are two pattern of GnRH secretion from hypothalamus: i) *Pulsatile secretion* which involves secretion of basal levels of the hormone from the tonic centre of the gland; ii) *Pre-ovulatory surge* which is associated with secretions from the surge centre of hypothalamus. This secretion prevents desensitization of GnRH receptors on the gonadotroph cells of the anterior pituitary. GnRH is transported to anterior pituitary *via* the hypophyseal portal blood system where it binds to G-protein coupled receptor on the cell surface of gonadotroph cells. This releases intracellular calcium which activates mitogen activated protein kinase (MAPK) signaling pathway leading finally to release of FSH and LH from storage compartments in the cytoplasm of gonadotrophs. FSH is stored in secretory granules for short periods of time whereas LH is stored for longer periods during the estrous cycle. These two hormones together drive the follicular phase of estrous cycle in presence of basal levels of progesterone due to regression of corpus luteum. The increased E2 concentration, from rapidly growing pre-ovulatory dominant follicle, together with decreasing progesterone concentration, induces a surge in GnRH and allows the display of behavioural estrus during which the female is sexually receptive. The pre-ovulatory GnRH surge induces a coincidental FSH and LH surge. When LH pulses occur every 40–70 min for 2 -3 days, the dominant follicle ovulates. This is followed by luteal phase which begins with met-estrus and is characterized by formation of corpus luteum from the collapsed ovulated follicle (corpus haemorragicum). The granulosa and theca cells of the ovulated follicle luteinize and produce progesterone for maintenance of pregnancy and/or resumption of the estrous cycle. Inhibin, produced by granulosa cells, suppresses production and/or secretion of FSH through negative feedback at pituitary level. The progesterone concentrations remain elevated during the following di-estrus phase. The recurrent waves of follicle development continue being initiated by release of FSH from anterior pituitary. The dominant follicles that grow during the luteal phase, however, do not ovulate primarily due to progesterone dominance which through negative feedback allows the secretion of greater amplitude but lesser frequency LH pulses that are inadequate for ovulation. If no fertilization occurs and/or embryo does not get implanted to uterine wall, corpus luteum starts regressing in response to prostaglandins secreted by the uterus. This causes decrease in progesterone concentration as a response to corpus luteum regression. The progesterone dominance vanishes which leads to increase in LH frequency and subsequent ovulation to continue the estrous cycle. If on the other hand, embryo gets implanted to uerine wall, progesterone dominance continues and further ovulation stops. The implanted uterus does not produce progesterone and hence, luteolysis of corpus luteum does not occur which finally leads to maintenance of pregnancy and cessation of estrous cyclicity. The animal enters anestrous (due to pregnancy) which continues through parturition and lactation (lactational estrous).

Artificial insemination

Successful artificial insemination (AI) depends on the quality of the semen, skill of the inseminator, timing of insemination and health status of the female. The optimal timing for

AI is 8–12 h after onset of estrus. The bio-stimulatory effect of a teaser bull leads to general increase in reproductive efficiency in a herd by improving the percentage of cycling animals. It improves the pregnancy rate from 19 to 43%. Although buffalo are polyestrous, fertility is reduced during the off-breeding season (spring to summer) and during the period of increasing daylight and summer heat.

Pregnancy and post-partum period

The gestation period of buffalo is longer than that of cow. It is approximately 310 to 330 days. Murrah has a shorter gestation period (315 days) than swamp buffalo (330 days). The calving interval of buffalo varies between 400 to 600 days, although longer calving intervals are also observed. The first ovulation in river buffalo generally occurs 55 days post-partum but may be delayed up to 90 days post partum when a suckling calf is present. In suckled buffalo the first estrous is detected 130 days post-partum. It may be delayed much longer depending on nutritional and climate conditions. Uterine involution is usually completed in 25–30 days. The stimulus of suckling shortens involution time. The period of post-partum anestrous is usually longer in buffalo as compared to cattle under comparative management conditions. Factors such as poor nutrition and body condition, suckling management and climate, nutrition, feed quality and availability influence the duration of this period. Post-partum anestrous is a major cause of infertility resulting in economic loss to buffalo breeders. LH secretion in buffalo remains low during early post-partum period and episodic pulses become detectable a few weeks before ovarian activity commences. Adequate nutrition before and after calving, restricted suckling and reducing heat stress by permitting wallowing or water sprinkling   reduce the duration of post-partum anestrous. The presence of buffalo bulls in a herd also has a bio-stimulatory effect which reduces irRegularities in estrous cycle pattern and advances the time of first post-partum ovulation.

Detection of estrus in buffalo

Buffalo is regarded as a shy or poor breeder. They are generally seasonally polyestrous, expressing heat for eight months of the year. A sexually inactive condition is observed from March to June, when hardly 3% of the heats are detected. The peak season of the estrus is from October to February. Indian buffalo show more conceptions when both diurnal temperature and relative humidity are low, i.e. in winter. The estrus symptoms in buffalo are less obvious than those in cattle. A less than a third of buffaloes in estrus can be detected by homosexual behavior. Symptoms such as swollen vulva, mucous discharge and increased frequency of urination are not regarded as reliable indicators of estrous. Silent estrus is a common problem in buffalo. The incidence of silent estrus is reported to be higher in herds using AI rather than natural service. This might indicate problems in estrus detection rather than the animal itself, necessitating the development of methodologies for efficient heat detection. A number of methods have been developed to overcome the problem of estrus detection which include but are not limited to:

1. Observation of symptoms: Though less obvious than in cattle, frequent urination, bellowing, vulvar swelling and mucus discharge serve as symptoms of an animal in heat. Acceptance of the male is considered as the most reliable estrus indicator in buffalo. Restlessness, moist, red, swollen and wrinkle-free vulvar lips, inappetitance, nervousness, riding or allowing to be ridden, reduction in milk yield, raised tail and crutching of back and lumbar region are other symptoms.

2. Presence of bull: If a bull is placed in a pen site in such a way that the buffaloes can pass closely and Regularly, the animals in estrous will generally migrate towards the bull.

3. Manual checking: Upon manual checking, the uterine horns are turgid and coiled with a more marked tone during estrus than during diestrum.

4. Bull parading: Parading of teaser or vasectomized bulls or androgenized buffalo bullocks in the shed at least twice during cooler part of the day is the common practice of heat detection in buffalo. The bull will sniff and mount on to the females which are in heat.

5. Chin ball marking device: Vasectomized bulls or cystic buffaloes can be used as heat detectors if fitted with chin ball marker. They will mark the backs of those buffaloes which they would mount. When the animal presses down with its chin on the back or rump region of mounted animals, a spring loaded valve in the device is opened and marking fluid is released on to the female in heat.

6. Painting of tail: A well-placed strip of paint on the tail head serves as a cheap and effective method. Paint will be rubbed off or at least cracked when the painted cow is mounted by another.

7. KaMaR heat mount detector: These devices are glued to hair over the middle just in front of the tail head. Pressure of mounting animal squeezes dye from the device and dyes the adjacent area.

8. Breeding record analysis: Probable date of incoming heat can be calculated by analyzing the breeding/ reproduction records of the individual buffalo.

9. Close observation: Close observation; say by use of close circuit television devices (CCTV) can be used to monitor the buffalo sheds.

10. Change in body temperature: Rise in vaginal temperature (by about 0.5 to 0.8 °C) during the estrus can be measured.

11. Pedometer: Pedometers are used to measure greater movement and activity of animals in estrus. The overall movement as well as activity is increased by 40% in estrus buffalo as compared to normal buffalo.

12. Vaginal probe: Lowest electrical conductivity of the fluid in the vagina at the time of estrus is the characteristic feature of using the probe.

13. Synchronization of the estrus: Artificial control of the estrus cycle by using progesterone or norgestomet –containing devices, progesterone releasing intravaginal devices (PRID), pessaries, ear implants along with PMSG (pregnant mare serum gonadotrophin), estradiol and or prostaglandin (PGF2α) have been used successfully to improve synchrony of estrus and conception in buffalo.

14. Trained dogs: Trained dogs for estrous detection can also be used.

15. Laboratory tests:

a) Progesterone assay: Milk progesterone level is lowered during the estrus period. It can be measured by ELISA or RIA and if the decreased level persists for 2 -3 days, the animal may be served. Progesterone assay could also be done in blood and serum.

b) Cervical mucus fern pattern test

c) Endometrial biopsy: Endometrial biopsy shows peak activity of phosphatases before start of estrus which persists for 1–2 h after onset of heat.

d) Vaginal smear: It shows an increase in cornified acidophilic cells during the heat period.

e) Viscosity of vaginal smear: The viscosity of vaginal smear is lowered during the heat.

However, caution should be applied for detection aids like heat mount detector, painting aids etc. which are unsatisfactory because wallowing or rubbing interfere with their efficiency and lead to false positive case detection.

Further reading

Mondal S, Prakash B and Palta P. (2007). Endocrine Aspects of Oestrous Cycle in Buffaloes (Bubalus bubalis): An Overview. Asian-Aust. J. Anim. Sci. 20: 124 – 131.

Drost M. (2007). Bubaline *versus* bovine reproduction. Theriogenology 68: 447–449.

Forde N, Beltman M, Lonergan P, Diskin M, Roche J, Crowe M. (2011). Oestrous cycles in Bos taurus cattle. Animal Reproduction Science 124: 163–169.

Perera B. (2011). Reproductive cycles of buffalo. Animal Reproduction Science 124: 194–199.

Suthar V. and Dhami A. (2010). Estrus detection methods in buffalo. Veterinary World 3: 94–96.

Lall HK. (1975). Study of economic characters in Murrah buffalo. Indian Veterinary Journal 52:337–344.

McIntosh J. (2008). The Ancient Indus Valley: New Perspectives. ABC-CLIO, Santa Barabara.

Mudgal VD. (1991). Nutritional programmes related to dairy buffalo production in developing countries. In: Proceedings of the Third World Buffalo Congress, Varna, Bulgaria 4: 962–977.

Mudgal VD. (1992). Buffalo meat. In: Encyclopedia of Food Science Technology and Nutrition. Academic Press, London, UK.p. 521.

Mudgal VD and Sethi RK. (1989). Riverine breeds of buffalo in Asia. In: Proceedings of a Seminar on Buffalo Genotypes for Small Farms in Asia. University Pertainian, Malaysia. pp. 27–44.

Chapter 4

Assisted Reproductive Techniques in Buffalo

Introduction

Buffalo, being the major source of milk and meat, significantly contributes to the economy of many tropical countries of the world, especially in south & south-east Asia, South America, Africa and the Mediterranean. But certain factors associated with reproduction like late maturity, poor estrus expression in summer season, low reproductive efficiency and fertility rates due to environmental stress coupled with poor management limit the efficient utilization of buffaloes in improving the living standards of farmers in these countries. Therefore, considerable attention has been focused to understand the causes responsible for limitations in reproduction among buffaloes by studying their physiology, reproductive endocrinology as well as by developing biotechniques to enhance their reproductive efficiency. The development of modern equipments like microscopy, centrifugation, ultrasonography, endoscopy, flow cytometry and sorting, microinjectors, micromanipulators, electroporators and nucleofectors, brought with them the era of novel, more improved, more efficient, more robust, more reliable and more reproducible techniques leading to the development of the era of assisted reproduction. Assisted Reproductive Techniques (ARTs), currently include, but are not limited to, artificial insemination (AI), multiple ovulation and embryo transfer (MOET), trans-vaginal ultra-sound guided oocyte recovery (TVOR), also known as Ovum pick up (OPU), *in vitro* fertilization (IVF), *in vitro* production of embryos (IVEP), cloning, cryopreservation of gametes and embryos, transgenesis, xenografting, germ-cell transplantation, pre-implantation genetic diagnosis, cryopreservation, sperm and embryo sexing. Its existence, sustenance and success pillars upon the way the buffaloes are/ and would be used for meat and milk production, for rapid multiplication of elite germ plasm, for germ plasm conservation, for multiplication of otherwise naturally incompetent but highly valuable animals, biopharming and the more recent and rightly much hyped endangered species (animal) conservation. Besides, its use in production, its importance lies in studying of reproductive processes to enhance the understanding and knowledge of the most mysterious and subtle part of life - "Reproduction," which is both the key to growth and backbone of animal economy. This chapter aims to provide an overview of assisted reproductive techniques and also of the research done to date towards the enhancement of buffalo reproductive efficiency through embryo biotechniques.

Historical background

The history of assisted reproduction may be traced back to seventeenth century when Anton Von Leeuwenhoek first observed the sperm under microscope and called them "animalcules." Thereafter in 1799, Lazzaro Splanzinni proposed that the contact between egg and sperm was necessary for the embryo to develop and grow. He further carried artificial insemination in dogs and succeeded in getting live births, and went on to inseminate frogs and fish, thereby establishing artificial insemination as the successful assisted reproductive technique. Spllanzini also led to the development of cryobiology with some of his early experiments where he succeeded in keeping the frog, stallion and human sperms viable after cooling in snow and re-warming. It was finally John Hunter who reported AI in humans by collecting the spermatozoa from a patient who suffered from hypospadias and injected it into his wife's vagina with a warm syringe, resulting in the birth of a child in 1785. Ivanoff reported AI in domestic farm animals, dogs, foxes, rabbits and poultry. He further developed semen extenders and initiated sperm freezing which subsequently lead to foundation for the establishment of artificial insemination as an animal breeding technique. Around the same time (1890s), the first successful mammalian embryo transfer was performed by Walter Heape by transferring two four-cell stage Angora rabbit embryos into an inseminated Belgian doe which subsequently led to birth of four Belgian and two Angora young. Thereafter, Alan Brachet succeeded in keeping alive the rabbit blastocyst in blood plasma for 48 hours. This was followed by establishment of successful pregnancies with embryos obtained after flushing from a number of species from mice and rabbits to sheep and cow. This embryo flushing and transfer became a routine in domestic animal breeding during the 1970s. Driven by the potential economic gains and commercial success, a range of assisted reproductive techniques have been developed for farm animal reproduction. Wilmut started initial works on animal cloning and set the stage ripe for animal science scientists to plunge into the era of cloning, by successfully producing Dolly- the first cloned animal in 1986. Thereafter, multiple assisted reproductive techniques have been established and their procedures standardized, to bring glory, name, fame and money to the science of assisted reproduction.

Drivers for Assisted reproduction

The main driver for the robust development of mammalian assisted reproductive techniques is the considerable plasticity of the mammalian oocyte and embryo. Mammalian oocyte is able to mature, face fertilization and develop into an embryo even in suboptimal conditions, as under *in vitro* environment. Such ability has encouraged, over the decades, the development of numerous *in vitro* assisted reproductive techniques in several species including humans. The use of ART has further increased because of its perceived safety both to the offspring and the mother. With the increasing intensity of infertility in humans (15% in reproductive age couples), coupled with the need of the deeper knowledge about pre-implantation development and embryo genetics, the evaluation of the overall ART safety has become a matter of great concern. This makes crucial the studies on pre-implantation embryo development in order to gain insights into the molecular mechanisms responsible for improved, efficient, safe and healthy assisted reproduction. Since, there is an obvious scarcity of human embryos for research; the use of appropriate animal models provides a reliable, cost- effective and irreplaceable support, thereby boosting the assisted reproductive intervention in farm animals, which considerably feature the characteristics of human fertilization. The economic gain and potential of commercial exploitation of these techniques in farm animals' added further oil to the wanting and haunting fire of humankind to drive the era of assisted animal reproduction into the boom.

Assisted reproductive techniques

Under current scenario, the assisted reproductive techniques are classified into four generations:

1. **First generation ART:** The first generation ART includes artificial insemination (AI) and gamete and embryo freezing.

2. **Second generation ART:** These include multiple ovulation and embryo transfer (MOET) technologies.

3. **Third generation ART:** These include *in vitro* fertilization (IVF) procedures. These techniques have matured into successful commercial applications, facilitating the increase in production through genetics, the reduction in generation intervals, the control of diseases and the cutback in production costs. Additional techniques that have evolved as different variants of IVF include gamete intra fallopian transfer (GIFT), zygote intrafallopian transfer (ZIFT), and intacytoplasmic sperm injection (ICSI).

4. **Fourth generation ART:** These include processes like cloning by nuclear transfer (NT) of embryonic and somatic cells, stem cell biology and transgenesis and such techniques which are still more experimental.

Despite of the generational classification, it becomes imperative to mention that all these techniques are intertwined and are completely inter-dependent. For example, we need cryopreserved spermatozoa (first generation ART) for IVF (third generation ART) to produce stem cells or to develop transgenic animals (fourth generation ART). It has been observed that the most consolidated reproductive techniques that have been genetically relevant in the past five decades involve mostly the first three generations of ART: primarily AI, cryopreservation of gametes and embryos, induction of multiple ovulations, ultrasonography, *in vitro* fertilization and embryo transfer. The third and fourth generation ARTs, have the potential of enhancing the influence of superior animals on production but their commercial adaptability, as of now, is limited. These ART include sexed semen or embryos, cloning, transgenics, stem cell biology and molecular diagnostics.

From AI to stem cell biology

Artificial Insemination: AI, the first generation ART, has been in use for the last 200 years. It has been the most successful and efficient reproduction technology in animal production for the last six decades, owing to successful achievements in semen cryobiology. AI has now become a practical technology in commercial dairy cattle programs in both developed as well as developing countries. The real revolution in AI came when frozen semen was used by Polge and his group (1949), which led to efforts to freeze buffalo semen until 1972. AI was initially used to spread improved indigenous breeds, which was followed by the more valuable crossbreeding. The main advantages of AI as an assisted reproductive technique in buffaloes are: rapid dissemination of superior genetic material, maximizing the use of superior males, greater efficiency and rate of genetic selection, introduction of new genetic material by import of semen rather by cumbersome import of the live animals, considerable reduction in transportation and maintenance hurdles and reduced risk of sexually transmitted diseases. It further enables animal science scientists to exploit the progeny tested bulls, even after their death, by using the frozen semen. Since, male animals are larger than females and thus require relatively more space and food, AI enables a farmer to save the costs required for maintenance of more strong, powerful and potentially ill-mannered male animals, thus saving the costs incurred in special housing and handling equipments. Such is the impact of AI that 50% of estimated increase in milk production in developed countries during the second half of the twentieth century, is attributed to genetic gain obtained by wide spread use of AI over conventional breeding.

AI was used for water buffalo in early forties at Allahabad Agricultural Institute in India and the first calf was born in 1943. Currently in India, 67 million frozen semen doses are produced and number of artificial inseminations performed has reached to 54 million, bringing about 22 million animals under AI with an overall conception rate of 35%. The government has planned to initiate a new scheme namely, National project on bovine breeding and dairy, which plans to bring 80% breedable females among cattle and buffalo under organized breeding through artificial insemination or natural service by bulls of high genetic merit. The project also has the mandate to arrange the AI services at the farmers' doorsteps.

Limitations: AI has some potential drawbacks that must be considered. AI is more laborious, the females which are in correct status for conception are not always detected in AI program, which are otherwise detected by the males instinctively. The prerequisite for AI is the identification of the best male, failure of which would reduce genetic gain as well as genetic variability in the population. Even, if the best male is used for AI the gene pool gets reduced due to production of more offspring per male. This prompts the necessity to balance the benefits of more intensive selection against the negative effects of decreased variation. Furthermore, in developing countries, the conception rate in field AI is very low, thus the desired effects in animal improvement have not been achieved. It is understood that

AI in this part of world will become effective only when considerably better technical and organizational facilities will be provided to farmers.

Gamete and embryo cryopreservation

It is rightly believed that the attainment of successful semen cryopreservation protocols sustained AI as the most successful and efficient ART. It was frozen semen that boosted the dairy industry by making AI simpler, economical, successful and widespread. One of the most critical requirements for *in vitro* embryo production (discussed ahead) is the continuous availability of viable and developmentally competent oocytes. Thus, considerable efforts were put in to devise the methodology which would enable the researchers to store and preserve the unfertilized oocytes without compromising their developmental potential. During the past few decades significant progress in oocyte cryopreservation has been achieved and live offspring of as many as 25 different species have resulted from transfer of cryopreserved oocytes or embryos. Like the preservation of its counterpart in AI, preservation of oocytes reduces the risk and expenses involved in live animal transport, hazards of diseases transmission and natural disasters and accidents. The preservation of oocytes from endangered species also safeguards them from extinction. The major problems *viz.*, mechanical damage and osmotic shock, which were main obstacles in gamete cryopreservation fell apart with the advent of such cryoprotectants like glycerol, ethylene glycol, propylene glycol (permeating cryoprotectants), sucrose, glucose or fructose (non-permeating agents). The process of vitrification was subsequently shown to be more effective than slow freezing for materials more sensitive to chilling. The cryopreservation of oocytes by vitrification was achieved in various species like bovine, swine, equine and buffalo. The successful embryo cryopreservation allowed the global commercialization of animals of high merit, as embryos. Embryo freezing is an established practice in commercial embryo transfer programmes where at least temporary storage of embryos, until they are transferred, is the essential need as the embryo viability starts declining after 12 hours in holding media. Moreover, the cryobanking of embryos is perceived as a very helpful strategy for conservation of endangered species of animals and thus demands an obvious attention owing to ruthless devastation of the natural habitats.

In vitro embryo production (IVEP)

The *in vitro* production of embryos in buffalo usually comprises of three steps: 1) *In vitro* maturation (IVM) of primary, germinal-vesicle stage, oocytes collected either directly from the ovaries of donor female animals (ovum pick up) or aspirated from the slaughter-house obtained ovaries; 2) *In vitro* fertilization (IVF) by combining IVM oocytes with *in vitro* capacitated sperm cells; and 3) *In vitro* culture (IVC) of presumed zygotes to developmental stages fit for transfer into recipient females. IVEP helps in production of high genetic merit animals, besides providing an excellent source of embryos for other assisted reproduction technologies like cloning, transgenesis, embryo sexing, stem cell research, etc. IVEP also provides sufficient embryos to be used for studying developmental and embryo genetics, proteomics, epigenetics as well as cytogenetic disorders. The speculation that early stages of bovine embryo development may show similarities with human embryos makes bovine embryos as model organisms and thus IVEP as a tool to generate the requisite raw material. The major thrust area in buffalo IVEP is to improve its efficiency which is currently around 30% to 40%, well below the expectations. Pratham was world's first buffalo calf produced through IVF at National Dairy Research Institute, India in 1990.

Despite several advantages of this technique, its initial application in cattle and buffaloes was limited, primarily due to scarcity of the oocytes from these species. However, the recent developments of transvaginal oocyte retrieval (TVOR) and ovum pick up (OPU) have removed these difficulties to a large extent. We have reported the success of this technique in giving birth to India's first female Sahiwal calf named 'Holi' from an aged animal. Also the standardization of procedures for buffalo IVM, IVF and IVC have been instrumental in thriving IVEP in poor breeding bubaline species. The greatest limitation of IVEP is the high production costs and the low overall efficiency under field conditions. Despite the hopes generated by IVEP, it remains still unclear whether IVEP provides a reliable alternative to conventional superovulation and embryo transfer for producing embryos from reproductively healthy buffaloes.

Semen and embryo sexing

Predetermination of the sex of offspring would inevitably lead to selective multiplication of the desired sex. It is presumed that known sex of embryos produced for use in ET programs can more effectively help to manage producer resources, by enabling to produce more heifer calves per ET, which is the main goal of the dairy entrepreneur. The presence and absence of Y elements determines the sex of embryo or semen and currently this is determined by: 1) Chromosomal analysis of demioocytes; 2) Immunological detection of embryonic H-Y antigen; 3) Fluorescent *in situ* hybridization; 4) Y-specific probes; 5) Loop mediated isothermal amplification reaction; and 5) Flow sorting of semen. It has been reported that bovine Y-specific sequences are conserved among buffalo, Indian *Zebu* and *Tarus* cattle. Thus, bovine Y-specific probes / primers could be used to demonstrate the sex of buffalo or zebu embryos. Embryo biopsy method is also used in which a single cell from early stage embryo is extracted and probed for presence or absence of Y chromosome. The offspring of desired sex have already been produced employing flow sorting for sexing of both fresh and frozen-thawed semen in several species like cattle, goat, pigs and sheep. The birth of the first buffalo calves produced by the combined use of AI and sexed semen has been reported by Presicce *et al.* (2005). The semen and embryo sexing has not been reported in field in any developing country except China. In India the refinement of the techniques of sperm and embryo sorting is currently limited to research institutions only.

Embryo transfer

Embryo transfer (ET) is a technique in which embryos are collected from a donor female and transferred to recipient females which serve as surrogate mothers for the remainder of pregnancy. It has been used to increase the reproductive performance of particular females of agriculturally important species like cattle, horse, sheep and goat. The commercial ET started way back in 1970s in North America primarily as a means of multiplying the number of exotic breeds of beef cattle. Then the advent of non surgical embryo recovery and transfer methods, lead to the expansion of this technique beyond the restricted domains of the surgical means of embryo recovery and transfer.

The initial success of embryo transfer in riverine buffalo in US was followed by the birth of calves in Bulgaria and India. In 1991, a riverine buffalo calf (2n= 50) was born out of transfer in a swamp buffalo recipient (2n= 48). The other crucial requirement for commercial exploitation was completed by the effective freeze-thaw method which permitted embryos to be shipped anywhere to the world. ET or MOET (multiple ovulation and embryo transfer) offers various commercial advantages in buffaloes like: i) easy and affordable transport of the desired embryos; ii) genetic improvement in domestic animal industry by obtaining a large number of desirable progeny from elite parents; iii) high quality breeding males to be available for sale; iv) exploitation of the developments in breeding technology like embryo sexing and embryo-splitting; v) utilization of the genetic contribution of the male and female at the same time; vii) production of artificial insemination sires from highly proven cows and bulls. The success of MOET program has lead to the use of this technology to test AI sires genetically. Under this program, selected buffaloes are super stimulated and inseminated to highly proven bulls. The male offspring are placed in waiting while female offspring are used for ET. The bulls are then proven by production records from siblings, rather than by progeny. Using this approach it is possible to test the bull genetically within three and a half years as compared to five and a half years using traditional progeny testing schemes. Pregnancy rates following non- surgical transfer of *in vivo* derived buffalo embryos was very low in early trials except for the 100% success achieved by Drost et al. in transferring a single embryo in 1983.

Although the basic procedures employed in buffalo for ET are established, considerable research is required for further improvement with the main focus on recovery of oocytes from live females by ultrasound guided aspiration and in subsequent IVF and IVC. Also at this time the technical costs involved in ET/MOET preclude its other applications but for seed-stock production, so re-consideration of the economics of ET is the need of the hour.

ET and the associated techniques have been utilized for the rapid multiplication of the elite breeds of cattle, buffalo, sheep, goat, horse and pig.

Cloning

Cloning represents the fourth generation of ART that has the potential to be used for multiplication of elite animals and minimize the genetic variation in the experimental animals, besides being used for conservation and propagation of endangered species. Somatic cell cloning also offers opportunities to select and propagate animals of specific merit or of desire. Cloning can also be used as a tool in therapeutics for generation of the autologous stem cells. Numerous types of cells have been used as donors in cloning, *viz.*, fetal fibroblasts, adult fibroblasts, granulosa cells, hepatocytes and lymphocytes. Wilmut *et al.* (1997) produced the first cloned animal (Dolly) and paved the way for the application of this assisted reproductive technique to almost every species like cattle, pig, goat and horse. A survey conducted by OIE in 2005 regarding the cloning technique, revealed that 91 countries, comprising especially the developing countries, have the cloning capabilities. Among the participating countries 60% were developing countries, 4% were from Africa, while 23% of the respondents were from Asia. Recently, the first cloned camel named, Injaz, a female, was born in 2009 and Bin Sougham, a male, was born in 2010 at the Camel Reproduction center in Dubai, United Arab emirates. The world's first buffalo calf named Garima was born at National Dairy Research Institute, India in 2009 by utilizing an in-house developed technique of Hand- Guided Cloning (HGC). Later on by utilizing the same technique, Garima-II and Lalima were also produced at the same institute, using buffalo embryonic stem cells and adult fibroblasts, respectively, as the donor cells. Garima II showed sound development and has given birth to female calf, Mahima, after AI. The institute also claims to have cloned an endangered Indian wild buffalo (Asha). The female calf, named Deepasha, was produced using the ear skin fibroblast of Asha as donor cell while demicytoplasts were isolated from slaughter house obtained ovaries.

Reproductive cloning holds the promise of bypassing the conventional breeding procedures to allow development of the duplicates of the genetically engineered animals. It is believed that cloning could be used in future in xenotransplantation to produce "humanized pig," the organs of which could be transplanted to humans. Cloning is also used efficiently for production of transgenic animals by employing several biotechnological techniques like pro-nuclear microinjection, cytoplasmic microinjection, retrovirus- and lentivirus- based vectors, nucleofectors, electroporators, etc. These transgenic founder animals, once produced, could be used both in breeding and biomedicine. There is currently a great interest in developing transgenic cows, buffaloes and goats producing recombinant proteins in mammary gland which could be ultimately harnessed through milk.

Stem cell technology and Transgenic buffalo production

An extensive work has been carried out in embryonic stem cell technology, spermatogonial stem cell technology and transgenic research in buffalo, especially at National Dairy Research Institute, Karnal. The culture systems have been developed for propagation and maintenance of buffalo embryonic stem cells. We recently reported the propagation of bubaline embryonic stem cells for more than 100 passages as well as the potential of these cells to differentiate into all three germ layers (ectoderm, mesoderm and endoderm). Signal transduction and different differentiation strategies have been studied in buffalo embryonic stem cells and culture systems have been developed to induce their differentiation into many cell types including germ cells. Spermatogonial stem cell culture strategies have been developed and the work for production of transgenic buffalo and goats is currently being carried out and it is hoped that transgenic animals secreting valuable human protein in milk will be a practical reality in a short span of time.

Future strategies

The procedures for IVEP viz., IVM, IVF and IVC have to be redesigned or modified to increase the efficiency of embryo formation in buffalo. This might be alleviated to some extent by introduction of sequential media, which would specifically suffice the nutritional and other requirements depending on the developmental stage. The placental incompatibility between the embryo and the surrogate mother has to be overcome in order to increase the efficiency of cloning and embryo transfer. The growing interest for producing transgenic founder animals would dominate the future pharmaceutical industries, and this consequently warns better understanding of SCNT cloning, in terms of both embryonic genetics and epigenetics. Embryo genomics has to be given the desired attention in order to elucidate the genetics of abnormal embryo production by cloning and other ARTs as well for understanding the embryonic defects at cellular level. One of our prime future strategies should be the introduction of stem cells and nanotechnology and their integration with other ARTs to produce such animals which may contain all desired characters.

Further reading

Betteridge KJ. (2003). A history of farm animal embryo transfer and some associated techniques. Animal Reproduction Science79: 203–244.

Campbell KHS, Fisher P, Chen WC, Choi I, Kelly RDW, Lee JH and Xhu J. (2007). Somatic cell nuclear transfer: Past, present and future perspectives. Theriogenology 68: S214–S231.

De Graaf SP, Evans G, Maxwell WM, Cran DG and O'Brien JK. (2007). Birth of offspring of pre-determined sex after artificial insemination of frozen thawed, sex-sorted and re-frozen-thawed ram spermatozoa. Theriogenology 67: 391–398.

Drost M, Wright JM, Cripe WS and Richter AR. (1983). Embryo transfer in water buffaloes (*Bubalis bubalis*). Theriogenology 20: 579–584.

Gajda B and Smorg Z. (2009). Oocytes and embryo cryopreservation-state of art and recent development in domestic animals. Journal of Animal and FeedSciences 18: 371–387.

Galli C and Lazzari G. (2008). The manipulation of gametes and embryos in farm animals. Reproduction in Domestic Animals43: 1–7.

MacKenzie A.A. Applications of genetic engineering for livestock and biotechnology products. Technical Item II, 73rd General Session, Paris, International Committee, OIE.

Presicce GA, Verberckmoes S, Senatore EM, Klinc P and Rath D. (2005). First established pregnancies in Mediterranean Italian buffaloes (*Bubalus bubalis*) following deposition of sexed spermatozoa near the utero tubal junction. Reproduction Domestic Animals 40: 73–75.

Saha A, Panda SK, Chauhan MS, Manik RS, Palta P and Singla SK. (2012). Birth of cloned calves from vitrified–warmed zona-free buffalo (*Bubalus bubalis*) embryos produced by hand-made cloning. Reproduction Fertility and Development 25: 860–865.

Shah SM, Neha S, Syma A, Zandi M, Manik RS, Singla SK, Palta P and Chauhan MS. (2015). Derivation, characterization and pluripotency analysis of buffalo ES cells derived from *in vitro* fertilized, parthenogenetic andhand-guided cloned blastocysts. Celular Reprogramming 17: 306–322.

Syed Mohmad Shah and Manmohan Singh Chauhan. (2015). Development of buffalo (Bubalus bubalis) embryonic stem cell lines from somatic cell nuclear transferred blastocysts. Stem cell Research 15: 633–639

Vlanov K, Karaivanov KH, Petrov M, Kachewa P , Alexiex A and Danew A. (1985). Studies on superovulation and embryo transfer in water buffaloes. In Proceedings of the First World Buffalo Congres3: 510.

Wilmut I, Schnieke AE, McWhir J, Kind AJ and Campbell KHS. (2007). Viable offspring derived from fetal and adult mammalian cells. Cloning and Stem Cells9: 3–7.

Zoheir KMA and Allam AA. (2010). A rapid method for sexing the bovine embryo. Animal Reproduction Science 119: 92–96.

Chapter 5

Recovering the Buffalo Oocytes

Introduction

Eversince the birth of first calf (Pratham) in buffalo following *in vitro* maturation and fertilization, the *in vitro* embryo production (IVEP) of buffalo has been gaining serious attention for research and commercial applications. The first pre-requisite for laboratory production of buffalo embryos is successful and efficient recovery of oocytes, either from slaughtered animal ovaries or from the live animal. Whatever the recovery method may be, availability of a large number of culture grade oocytes is an essential criterion to realize a greater number of pre-implantation embryos under laboratory conditions. This becomes more important considering the fact the buffalo responds very poorly to superovulation treatments as compared to cattle. Thus the interest in recovery of oocytes from unstimulated ovaries has increased in buffalo for *in vitro* embryo production schemes. The success of *in vitro* embryo production has further hampered due to several factors like lesser number of follicles on the ovaries, poor oocyte recovery, anestrus associated with ovarian acyclicity and poor recruitment of primordial follicles into growing and Graafian follicles. Considering these intrinsic limitations associated with buffalo, an intense need was felt to develop efficient methodologies for recovery of the maximum possible number of oocytes from the ovaries. With the slaughter of more than 2 million buffaloes annually in India, the ovaries obtained from slaughter houses provide the cheapest and most abundant source of primary oocytes. A number of oocyte recovery techniques have been developed, most common of which are aspiration, slicing and follicle puncture. A number of techniques have also been developed for collection of oocytes from a live animal, whether stimulated or unstimulated. The important once are ovum pick up, laproscopic aspiration as well as endoscopic collection through right paralumbar fossa.

Methods for oocyte retrieval

A) Slaughtered animals

1. Follicular aspiration

The aspiration method is commonly employed for collection of oocytes from abattoir obtained buffalo ovaries because of the convenience associated with its application. Aspiration of the oocytes is done using a needle attached to a 10 ml syringe. A 16–18G syringe needle is most commonly used in order to avoid disruption of the surrounding cumulus cells. Pipette and aspiration needle under vacuum pressure have also been used for the purpose. It is recommended to use plastic disposable syringes for the purpose because there is possibility of toxicity with glass syringes containing rubber plungers and siloxane lubricants. If such syringes are used it is advised to wash and sterilize them under stringent conditions as employed for tissue culture glassware. If aspiration is performed by aspiration needle under a vacuum pump 16–18G needle attached to a vacuum pump under pressure of 75–100 mmHg is used. An increase in vacuum pressure reduces the number of viable oocytes, presumably because of disruption of cumulus cells. However, to maximize the number of good quality oocytes, an aspiration vacuum pressure of <50 mmHg is recommended. Aspiration of the visible follicles (2–6 mm diameter) on the surface of the ovaries is done by inserting the needle tip attached to the syringe containing 1–2 ml of the oocyte collection medium (TCM 199 + 50 µg/ ml Gentamicin sulphate) into the follicle from the side followed by slow outward movement of the plunger with the thumb. A number of ovaries are aspirated in similar manner till the syringe gets filled. The contents of the syringe (follicular fluid + cumulus oocyte complexes) are then collected in a 15 or 50 ml plastic test tube and kept undisturbed in a thermostat at 38˚C. After about half an hour when the debris settles down, the supernatant is removed gently and the sediment is laid onto the intergrid plates and mixed with additional oocyte collection medium. The cumulus oocyte complexes (COCs) are then searched from the debris under 20X zoom stereomicroscope. The COCs are washed thrice in the oocyte washing medium and finally the good quality COCs are put for *in vitro* maturation or used for other purposes, as needed by the researcher (Figure 5.1).

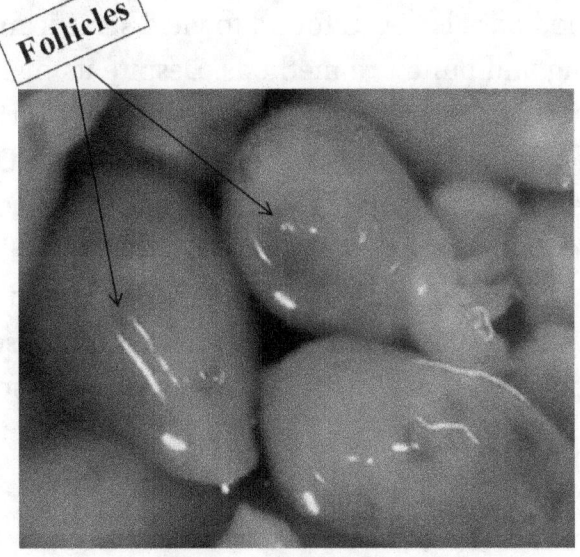

Fig. 5.1: A) **Buffalo ovaries**

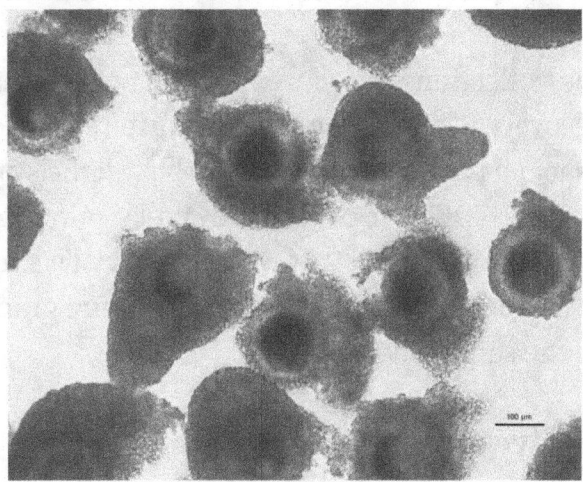

Fig. 5.1: B) **Aspirated buffalo oocytes**

2. Slicing of the ovaries

In this method, an ovary is held firmly in position by a hemostat attached to its base and 2–3 mm deep incisions are made across the whole ovarian surface using a sterile scalpel blade. The cut ovary is then swirled vigorously in a beaker containing oocyte collection medium. The medium is then searched for COCs as done in aspiration. Slicing techniques are either applied directly to the ovary or applied after aspiration. Slicing retrieves oocytes from both the follicles under the ovarian surface (cortical location) as well as peripherally located follicles (surface-visible), unlike aspiration which retrieves surface follicles only.

3. Puncture of ovarian surface

This method involves puncture of whole ovarian surface using a sterile 18G hypodermic needle while the ovary is held completely submerged in oocyte collection medium in a 90 mm Petridish. The medium is then searched for COCs under a 20X zoom stereomicroscope.

Comparison of the oocyte retrieval methods

Recovery of oocytes by aspiration of antral follicles has been the most commonly employed method. The main advantage of follicular aspiration is in terms of speed of operation which is particularly important for *in vitro* embryo production. Aspiration is perceived to be three times faster than slicing but the main disadvantage is the lower yield of oocytes as compared to slicing. This is primarily due to the fact that oocytes are recovered from 30–60% of the punctured follicles which is far lesser than almost 100% recovery in slicing. Though slicing takes three times longer for an ovary than aspiration, this is compensated for by the threefold greater yield. Thus in terms of workload, slicing provides the same number of oocytes per unit of time but with fewer ovaries. Aspiration method has also been found to yield significantly lesser number of good quality COCs than slicing and puncture methods. Despite its higher yield, slicing is open to criticism on the grounds of risking contamination of the COCs during the collection procedure. There have been different reports for the mean number of COCs recovered from an ovary using different recovery methods. Though no agreement has been reached for the actual number but all the studies indicate that slicing yields highest number of total COCs per ovary followed by puncture, while aspiration the lowest. We retrieved an average of 7, 4 and 2.5 COCs per ovary by slicing, puncture and aspiration (no vacuum used) methods, respectively. The mean culture grade oocytes were 5, 2 and 1.5 for the three methods in the same order.

B) Live animals

Ultrasoung-guided transvaginal oocyte retrieval

The most successful and most commonly used method for oocyte retrieval from a live buffalo female is ultrasoung-guided transvaginal oocyte retrieval (TVOR), commonly referred to as ovum pick up (OPU). The retrieval of oocytes from abattoir obtained ovaries has provided an alternative to utilize the female gamete pool, just like artificial insemination has provided for the male gamete pool. However, the oocytes collected from slaughter-house ovaries suffer from serious limitation of having unknown pedigree of the offspring born. The pool is further heterogenous and is mostly from low quality animals which are given to slaughter. For producing offspring of known genetic background, oocytes have to be obtained from live animals of known pedigree and for this OPU is the most used technique. The little genetic value and anonymous pedigree of abattoir-derived oocytes has led to increased adoption of OPU in both cattle as well as buffalo as an alternative to superovulation as well as an opportunity to increase the maternal contribution to genetic improvement. The linking of this low invasive oocyte-retrieval approach with *in vitro* embryo approach proved to be a new valuable development in breeding and embryo transfer industry. OPU enables the repeated collection of oocytes from live animals on a weekly or bi-weekly basis over long periods of time without causing any harm to animal.

Procedure

OPU involves the puncture and aspiration of ovarian follicles from live animals with the help of a needle attached to a vacuum pump which is guided ultrasonically to the ovarian follicle through the vaginal wall. The oocyte retrieval is performed under general (Rompun) and epidural (2% Xylocaine) anaesthesia, administered 15 min prior to the start, in case of buffalo. We performed follicular aspiration using a ultrasound machine (Aloka SSD-500, Japan) with a 5 MHz transvaginal convex transducer, equipped with a needle guide, single lumen 18G 55cm long sterile needle with an ultrasound echo tip and a vacuum pressure of 90 mmHg. A constant pressure of 50 to 90 mmHg is applied till the follicle gets disappeared from the monitor. This results in aspiration of the follicular contents into the needle and collapsing of the follicle which can be viewed on the monitor screen. The procedure is repeated for the adjacent follicle and for other follicles in other ovary. Following this the needle is withdrawn under continual suction and flushed with the flushing medium (DPBS⁻ supplemented with 50 µg/ml gentamicin, 20 µg/ml heparin and 0.3% lyophilized bovine serum albumin). The contents (oocytes, follicular fluid, granulosa cells and tissue debris) retrieved were allowed to settle down for 10–15 min in 50 ml falcon tube. The sediment was collected and searched for oocytes under a zoom stereomicroscope at 20X magnification. The collected oocytes were graded and subjected to *in vitro* maturation, fertilization and culture (IVMFC) to produce embryos which were then transferred to suitably synchronized recipients for production of offspring of known pedigree. It is important to note that the size of follicles is determined before puncture. The follicles are characterized on the basis of their diameter into small (3–5 mm), medium (6–9 mm) and large (≥10 mm). Most commonly medium and large follicles are used for aspiration while the small follicles are not punctured.

The number of oocytes retrieved, oocyte quality and embryonic development rates following IVMFC (*in vitro* maturation, fertilization and culture) is influenced by several factors like frequency of OPU, season and hormonal treatment for superovulation. It has been observed that twice a week frequency results in a significantly higher number of small, medium and large follicles per buffalo per week as compared to once a week aspiration. We observed greater number of grade A, grade B and grade C oocytes in twice a week OPU frequency than once a week. Twice aspiration per week further resulted to higher number of culturable oocytes (grade A and B), cleavage rate, 8-cell, 16-cell and transferable embryos per buffalo per week when compared to once a week aspiration. The number of follicle, oocytes recovered and embryos developed were higher in peak breeding season as compared to off-breeding season. Aspiration twice a week during peak breeding season resulted to a significantly higher number of small, medium and total follicles recorded as compared to aspiration once a week. The breeding season, however, was found not to significantly affect the number of large follicles and oocyte recovery rate. The maximum number of oocytes retrieved per buffalo per week was 2.75 when OPU was conducted twice a week during the peak breeding season. Among the gonadotropins used, follicular development was found to be greater in response to FSH than to PMSG. The mean number of oocytes recovered per donor was higher for FSH treated group as compared to PMSG treated group. However, no difference in oocyte type (mature or immature) was seen between FSH and PMSG treated animals.

The OPU approach assumes more significance in buffalo because superovulation has not shown success in buffalo, primarily because of poor responses to superovulation protocols involving exogenous follicle stimulating hormone (FSH) treatment. Despite a number of studies, employing different superovulation protocols conducted in this species, the number of viable embryos per flush remains very less as compared to cow (1.4 versus 5). Thus, OPU in unstimulated buffalo cow provides the only alternative for production of a large number of pedigree analyzed and genetically valuable buffalo embryos for commercial or research purposes.

Grading of the retrieved oocytes

Oocyte quality is the result of multiple factors that should be carefully taken into account for both *in vitro* and *in vivo* embryo production. A number of approaches have been developed to analyze oocyte quality that focus on ovary, follicle, cumulus-oocyte complex and the oocyte. The morphological evaluation of the cumulus-oocyte complex is the most common procedure used to assess the oocyte quality but the clear correspondence between visual criteria and developmental competence of oocytes under *in vitro* embryo production are yet to be established. Since, the quality of the raw material strongly affects the production efficiency; a good quality oocyte is the pre-requisite for successful *in vitro* embryo production. This is more important when oocyte population is from a heterogenous source like abattoir-derived ovaries.

Determinants of oocyte quality

The quality of an oocyte from a functional point of view refers to its developmental competence, i.e., capability to form an embryo after fertilization. Five levels of oocyte competence are generally described: 1) ability to resume meiosis; 2) ability to undergo cleavage after fertilization; 3) ability to develop into blastocyst; 4) ability to induce pregnancy and bring it to term; 5) ability to develop to term in a good health. These abilities actually follow a chain of separate but interrelated events. The ability to succeed in the first event does not ensure success of the subsequent event. Since, functional evaluation is difficult to be performed at the time of oocyte collection so different strategies were employed to provide a predictive value of *in vitro* embryo production of the collected oocytes. Most of the evaluations are based on the morphological criteria of the cumulus oocyte complex which are assumed to be related to the physiological status of the follicle and developmental potential of the oocyte. On the basis of visual assessment of morphological features observable under a light microscope, oocyte (bovine as well as buffalo) quality is assessed on the basis of following criteria:

1. A- grade: The oocytes in this grade have compact multilayered (> 5 layers) cumulus cells the ooplasm is homogenous and the COC is light and transparent.

2. B-grade: The oocytes have multilayered and unexpanded cumulus investment (≥ 5), homogenous ooplasm but with a dark zone at the periphery of the oocyte. The COC is slightly darker and less transparent.

3. C- Grade: The cumulus layers are fewer or the oocyte is naked and the ooplasm is irRegular with dark clusters. COC is darker than A and B grade.

4. D-grade: The cumulus layers are expanded and the cumulus cells are scattered in dark clumps in a jelly matrix. The ooplasm is irRegular with dark clusters. The COC is dark and irRegular.

The COCs of A and B grade are usually selected for *in vitro* embryo production while those of C and D-grade are rejected as their developmental competence has been found to be very low (Figure 5.2).

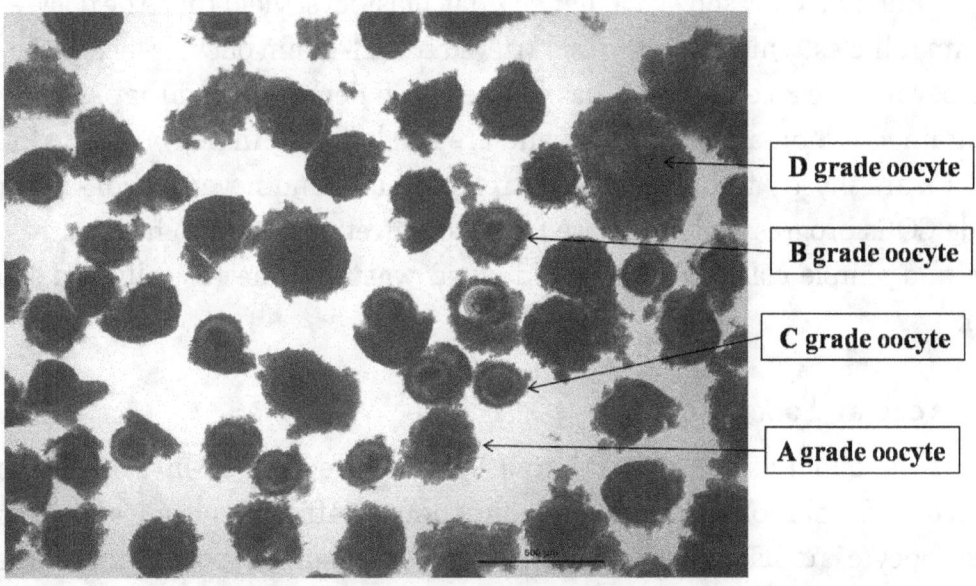

D grade oocyte

B grade oocyte

C grade oocyte

A grade oocyte

Fig. 5.2: Different grades of aspirated buffalo oocytes

Factors affecting oocyte quality

A number of factors affect oocyte quality like age of the animal, stage of the estrous cycle, hormonal patterns and biochemical characteristics of follicular fluid, diameter of the follicle from which oocyte is retrieved, atresia grade of the follicle, ovarian morphology and reproductive status of the donor. It has been reported that pregnant cows yield a higher number of high-quality oocytes than cyclic animals, probably due to higher progesterone levels in circulation and constant follicular turnover.

A) Ovarian features and oocyte quality

Oocyte quality is a function of follicular dynamics and, hence, depends on several factors like environment, seasonal variation, heat stress, physiological status, estrous cycle activity and genetics. A comparison of oocytes collected from early luteal, late luteal, follicular phase and non-cyclic cow ovaries, revealed that oocytes from early luteal phase ovaries exhibited a greater developmental competence in terms of the number of blastocysts formed as compared to others. It has also been shown that the bovine ovaries with lesser than 10 number of 2 to 5 mm diameter follicles and no follicle > 10 mm yield oocytes that show lowest *in vitro* embryo production efficiency and yield blastocysts with lowest number of cells in contrast to ovaries with follicle greater than 10 mm diameter or more than 10 follicles of 2 to 5 mm diameter. Though the exact reasons are not known but systemic factors like diet, environment, age, season, or individual variation of growth factors which mediate ovarian stimuli, might be the possible reasons.

B) Follicle size and oocyte quality

A positive relationship has been demonstrated between follicle size and developmental competence of the oocyte. It was found in bovine that the oocytes retrieved from larger follicles (4–8 mm diameter) showed a higher 7-day blastocyst yield than the oocytes retrieved from 2–4 mm follicles, while the oocytes retrieved from 1–2 mm oocytes lacked the capability to develop beyond the 8-cell stage. It has further been demonstrated that A-grade COCs are mostly contained in non-atretic follicles and are rarely found in heavy-atretic follicles. The C-grade COCs are mostly found in heavy-atretic follicles and never in non-atretic follicles. The B-grade COCs progressively increase from non-atretic follicles to heavy-atretic follicles. This gross and simple classification would avoid wasting time for follicle dissection and evaluation.

C) Cumulus cell and oocyte quality

The study of the cumulus cell transcriptome profile offers a non-invasive opportunity to predict oocyte and embryo competence as bidirectional traffic of molecules between cumulus cells and the oocyte is crucial for acquisition of this competence. A number of candidate genes expressed in cumulus cells could be used as indirect valuable markers of oocyte competence and quality. A set of genes like hyaluronan synthase 2 (HAS2), inhibin βA (INHBA), epidermal growth factor receptor (EGFR), gremlin 1 (GREM1), betacellulin (BTC), CD44, tumor necrosis factor-induced protein 6 (TNFAIP6) and prostaglandin-endoperoxide synthase 2 (PTGS2) could be used as potential biomarkers of oocyte quality.

A number of oocyte characters may also be used for evaluation of oocyte quality but unfortunately most of these evaluation procedures require the removal of the cumulus layers or even more invasive procedures that are not compatible with the viability and developmental competence of the oocyte, hence, cannot demonstrate the truthfulness of the analysis. A number of studies have been conducted to relate oocyte diameter, distribution of organelles like mitochondria and cortical granules, plasma membrane electrical properties and intracytoplasmic calcium stores to oocyte quality and developmental competence. The oocyte is an electrogenic cell and is capable of responding to electrical stimuli and modifying its electrical properties during the crucial periods of maturation and fertilization. Thus, the mitochondrial activity, the distribution of cortical granules and ion channel activity manifest discretely upon the oocyte quality. The modification of intracellular calcium levels in gametes has been extensively studied, and these modifications are recognized to be second messenger system for gamete maturation and fertilization. Calcium stores have been shown to bear a relation to morphological quality in immature oocytes and to developmental competence in *in vitro* matured oocytes.

Further reading

Assidi M, Dufort I, Ali A, Hamel M, Algriany O, Dieleman S and Sirard MA. (2008). Identification of potential markers of oocyte competence expressed in bovine cumulus cells matured with follicle-stimulating hormone and/or phorbol myristate acetate *in vitro*. Biology of Reproduction, 79:209–222.

Blondin P and Sirard MA. (1995). Oocyte and follicular morphology as determining characteristics for developmental competence in bovine oocytes. Molecular Reproduction Development, 41:54–62.

Boni R, Cuomo A and Tosti E. (2002). Developmental potential in bovine oocytes is related to cumulus-oocyte complex (COC) grade, calcium current activity and calcium stores. Biology of Reproduction, 66:836–842.

Tosti E and Boni R. (2004). Electrical events during gamete maturation and fertilisation in animals and human. Human Reproduction Update, 10:1–13.

R Boni. (2012). Origins and effects of oocyte quality in cattle. Animal Reproduction, 9, 333–340.

M. Techakumphu T, Lohachit C, Tantasuparak I, Intaramongkol C and Intaramongk S. (2000). Ovarian responses and oocyte recovery inprepubertal swamp buffalo (bubalus bubalis) calves after FSH or PMSG treatment. Theriogenology 54:305–312.

Mutha R and Uma M. (2012). Efficacy of different harvesting techniques on oocyte retrieval from buffalo ovaries. Buffalo Bulletin, 31: 209–213.

Samad H and Raza A. (1999). Factors affecting recovery of buffalo follicular oocytes. Pakistan VeterinaryJournal, 19: 56–59.

Manik RS, Singla SK, Palta P and Madan ML. (1998). Ovarian follicular dynamics monitored by realtime ultrasonography during oestrous cycle in buffalo (Bubalus bubalis). AsianAustralasian Journal of Animal Sciences, 11:480–485.

Appendix 1

Aspiration and classification of oocytes

We employed the following protocol for oocyte aspiration and their grading in our laboratory.

Buffalo ovaries collected from Delhi abattoir, immediately after slaughter, were transported to the laboratory in warm antibiotic (400 IUml^{-1} penicillin and 500 µgml^{-1} streptomycin) fortified saline within 3–4 h. In the laboratory, the ovaries were rinsed twice, trimmed to remove extra-ovarian tissue and washed properly (5–6 times) with warm saline containing antibiotics.

Oocytes were collected by aspiration of surface follicles (2–8 mm diameter) with an 18 gauge needle attached to a 10 ml syringe, containing aspiration medium (TCM-199 + 2.0 mM L-Glutamine + 0.3% BSA + 50 mg ml^{-1} gentamicin sulfate). The contents of the syringe, which included the aspirated cumulus-oocytes complexes (COCs), follicular fluid, granulosa cells and other debris, were poured in 100 mm x 100 mm square Petri dishes with 13 mm grid to be searched for the COCs under a zoom stereomicroscope at around 20X magnification. The searched oocytes were collected in 35 mm Petri dishes containing the washing medium (TCM-199 + 10% FBS + 0.81 mM sodium pyruate + 2.0 mM L-Glutamine +50 mgml^{-1} gentamicin sulfate) and washed twice in the same medium. The aspirated oocytes were graded according to the following criteria:

Usable quality: Compact cumulus oocyte complexes (COCs) with an unexpanded cumulus mass having ≥3 layers of cumulus cells, and with homogenous, evenly granular ooplasm.

Unusable quality: COCs partially or wholly denuded or with expanded or scattered cumulus cells or with an irRegular ooplasm.

Only usable quality oocytes were processed for *in vitro* maturation.

Normal saline containing antibiotics for ovary transportation	
Composition	Volume (1000ml)
Sodium chloride	9.0 gm
Penicillin G/streptomycin	0.06 gm/0.1gm
Distilled water	1000 ml

Aspiration medium (for about 200–250 ovaries)	
Composition	**Volume (50 ml)**
TCM-199 (Hepes modified)	50 ml
BSA	0.15 gm
Gentamicin	50µg/ml
L-glutamine	0.68mM

Washing medium	
Composition	**Volume (40 ml)**
TCM-199 (Hepes modified)	36 ml
FBS	4 ml
Sodium pyruvate	0.0036 gm
Gentamicin	50µg/ml
L-glutamine	0.68mM

Chapter 6

In Vitro Maturation of Buffalo Oocyte

Introduction

Sexual reproduction is completely dependent upon meiosis for generation of haploid gametes. Despite this universal requirement across all the mammalian species, meiosis is regulated differently in oocytes and spermatocytes. The spermatocytes start meiosis at puberty and proceed through the meiotic divisions uninterrupted, whereas oocytes get invariably arrested during one, and sometimes two times at different stages of the meiosis process. The oocytes overcome first meiotic arrest in response to intercellular signaling in a process called oocyte/meiotic maturation. Oocyte maturation also enables an oocyte to acquire fertilization competence and embryonic developmental potential. During the maturation process, oocyte must construct and store excessive number of Golgi, mitochondria, ribosomes, specific sperm receptors and nutritional reserves. The oocyte also prepares itself to ensure productive fertilization since aberrations in fertilization, like polyspermy, result in early death of the embryo. Given the importance of correct oocyte maturation for successful fertilization and subsequent embryonic development, the process of oocyte maturation assumes a remarkable importance for *in vitro* embryonic production. Since, under *in vitro* embryo production, the oocytes are aspirated from pre-ovulatory follicles at an immature stage; the correct and efficient maturation becomes the first important criterion for successful embryo production, whether through *in vitro* fertilization, parthenogenesis or somatic cell nuclear transfer. Before going into the details of *in vitro* oocyte maturation, it becomes imperative to know how the process is accomplished under *in vivo* conditions.

Oocyte maturation

Once the ovary is assembled from somatic and germ cells, the primordial germ cells proliferate to populate the gonad, start meiosis and finally differentiate to oocytes. But the oocytes are arrested in prophase I of meiosis during the fetal period. The completion of the first meiotic division takes place when oocytes have undergone extensive growth in cellular interaction with granulosa and theca cells. The meiotic process proceeds up to diplotene stage of meiosis I only and gets arrested until puberty in domestic animals. The release of this oocyte from the arrest, release of first polar body and its progression from the diplotene to metaphase stage of meiosis II (diakinesis) constitutes oocyte maturation. It is a complex process and involves cytoplasmic and morphological changes in the oocyte in addition to diakinesis. The oocyte resumes meiosis in response to LH (luteotrophic hormone) surge but gets arrested again at metaphase II stage and remains so until ovulation. In farm animals, the arrest is overcome by fertilization with a sperm which leads to release of a second polar body. In most animals, the growing oocyte contains an enlarged nucleus with a large nucleolous, called as Germinal vesicle (GV). GV is transcriptionally very active and mostly contains lampbrush chromosomes in animals with large oocytes. It transcribes an extensive array of genes whose products are necessary for oocyte development and for sustenance of early embryonic development. The oocyte accumulates an extensive collection of RNAs, proteins and organelles, such as cortical granules, yolk vesicles, ribosomes and mitochondria. With the initiation of maturation various nuclear and cytoplasmic changes begin within the oocyte enabling it to resume meiosis and achieve fertilization capability.

Nuclear maturation

Oocyte nuclear maturation refers to all the processes that lead to progression of its nucleus from the germinal vesicle to metaphase II stage. It involves Germinal Vesicle Breakdown (GVBD), chromosome condensation, metaphase I spindle formation, separation of homologous chromosomes, extrusion of first polar body and arrest at metaphase II. The most striking event in oocyte maturation is GVBD which begins with undulations of the nuclear envelope and chromosome condensation. Breaks in the nuclear envelope can be detected within 2 h, and after approximately 3 h the nuclear envelope completely disappears in rabbit, rat and mouse oocytes, while in humans and animals this process takes 20–24 h. During GVBD and chromosome condensation, kinetochores and the microtubule system appear to organize the spindle formation. The spindle apparatus increases in size and moves to the periphery of the oocyte. Metaphase I lasts for few hours and leads to anaphase and telophase when homologous chromosomes become separated and one half is extruded with cytoplasmic material such as mitochondria, ribosomes and cortical granules into the perivitelline space. The division proceeds into meiosis II to get arrested at metaphase II which marks the beginning of fertilization or parthenogenetic activation of the oocyte. Nuclear maturation involves changes in protein synthesis patterns. Bovine and buffalo oocytes undergo marked changes in the patterns of protein synthesis after GVBD. The ability of an oocyte to complete meiosis is known as meiotic competence and growing oocytes can be categorized as incompetent or competent depending upon their size and Cyclin B levels. Incompetent oocytes remain at GV stage only as they do not possess enough Cyclin B to progress beyond prophase I. These oocytes have lesser diameter (110 μm in case of bovines) and grow in antral follicles whose diameter remains less than 2 to 3 mm. The meiotic competent oocytes, on the other hand, have diameter >110 μm, and their follicles reach 2–4 mm diameter. The oocytes possess high levels of Cyclin B, p34 and cdc25 which specifically dephosphorylate the specific threonine and tyrosine residues of p34.

Bovine oocytes with an inside zona diameter smaller than 95 μm are unable to resume meiosis *in vitro*. A high proportion of bovine oocytes are able to resume meiosis to the MI stage once the diameter is at least 100 μm. The oocyte, however, must reach 110 μm or more to reach MII stage. Cleavage and blastocyst rate bears a positive correlation with meiotic competence and blastocyst formation rate, following *in vitro* fertilization.

Cytoplasmic maturation

Cytoplasmic maturation refers to ultrastructural changes that occur in oocyte cytoplasm and acquisition of developmental competence. It is assessed as the ability of the mature oocyte to undergo normal fertilization, cleavage and blastocyst development. The morphological parameters associated with cytoplasmic maturation include cumulus cell expansion, expulsion of first polar body and an increased perivitelline space. The organelle position and shape within the ooplasm bears a relation to oocyte size. For example, GV is located eccentrically in bovine oocytes with > 110 μm diameter, while those with diameter < 110 μm have GV located close to zona pellucida. As the oocye matures it shows an increase in its size as well as decrease in transcription rate. Nucleolous inactivation occurs during the growth of bovine oocyte from about 110 to 120 μm. Mitochondrial shape and distribution also change with oocyte size, as the oocyte matures. Mitochondria are round and centrally located in oocytes with > 100 μm diameter, while they are hooded and located peripherally in oocytes with <110 μm diameter. The change in location of cortical granules constitutes the most obvious ultrastructural sign of cytoplasmic maturation. These are originally located at center of oocyte but translocate to periphery and finally become attached to plasma membrane. The pattern and location of cortical granules seems crucial for normal fertilization as the release of contents from these granules changes the physical and chemical properties of zona pellucida, in addition to increasing the perivitelline space and prevention of polyspermy.

Regulators of oocyte maturation

The pre-ovulatory ovarian follicle, composed of an oocyte and surrounding theca, granulosa and cumulus cells, interact with each other to create a symphony of cellular signaling which finally leads to oocyte maturation and subsequent ovulation. The ovulated mature mammalian oocyte is a truly unique cell in female's for being the largest, though transient, cell in the body, being surrounded by its own extracellular matrix (zona pellucida) and having an arrested haploid nucleus that is transcriptionally inactive. Oocyte maturation is regulated *in vivo* by a complex interaction between oocyte, cumulus cells, ooplasm, gonadotrophins, growth factors, steroids and ions. Since, maturation is a much delayed process lasting for a couple of years in domestic animals; these factors interact in a complex manner to sustain the oocyte in diakinesis. The role of each of these factors in maturation process is briefly discussed.

1. Cumulus cells

The most remarkable feature of an oocyte is its intimate relationship with the layer of specialized somatic cells, cumulus cells, surrounding it. These cumulus cells are connected to the oocyte through a complex network of cytoplasmic extensions and gap junctions, forming a germ-somatic cell syncytium in the follicle. This places cumulus cells at the signaling nexus between the oocyte, the rest of the follicle and thereby with the endocrine system. Luteinizing hormone (LH) is understood to be involved precisely in final stages of oocyte maturation. Though the hormone acts *via* multiple signaling pathways, it has been established that LH receptors are present on theca, mural granulosa cells as well as on cumulus cells (in some species only), but not on the oocyte itself. This further demonstrates the importance of oocyte-somatic cell association for oocyte maturation. Granulosa cells also produce C-type natriuretic peptide (CNP) that finally leads to enhanced cGMP (cyclic Guanosine monophosphate) production in these cells. The resulting cGMP is then transferred to the oocyte *via* gap junctions which maintain oocyte meiotic arrest.

2. Luteinizing hormone (LH)

Maturation events like release from diakinesis, pre-ovulatory follicle rupture and release of fertilization-competent cumulus-oocyte complex (COC) are triggered by rapid increase in circulating levels of LH, secreted by the anterior pituitary gland. Though LH signaling assumes crucial importance for oocyte maturation process, the complete signaling mechanism is yet to be known, especially in farm animals.

It is established that the elevated levels of cAMP (cyclic adenosine monophosphate) in ooplam are responsible for meiotic arrest of an oocyte. It was known for a long time that cumulus cells produce cAMP which then enters into oocyte by gap-junctional coupling between the cumulus layer and the oocyte. Recent studies in mice have shown that cAMP is produced within the oocyte and is maintained primarily by constitutively active G-protein coupled receptors (GPR3 and GPR12) which stimulate adenylate cyclase activity. The increased levels of cAMP maintain activity of protein kinase A (PKA) which in turn prevents the activity of maturation promoting factor (MPF), thereby preventing entry into M-phase. Thus, resumption of meiosis should be associated with a decrease in oocyte cAMP brought about by LH surge. The situation is, however, contradictory in that LH surge causes an increase in follicular and oocyte cAMP levels and not the expected decrease. Therefore, some other mechanism seems to be involved in oocyte maturation following LH surge. It has been established that LH surge activates a powerful phosphodiesterase activity within the oocyte, kept otherwise in check. The elevated phosphodiesterase activity, especially by PDE3A, rapidly hydrolyzes cAMP and leads to meiotic resumption. The activity of phosphodiesterases in inhibited during oocyte diakinesis by elevated cGMP levels within the ooplasm. The cGMP level is maintained by a membrane-bound guanylate cyclase (mGC) on granulosa and cumulus cells which is then transferred into ooplasm *via* gap junctions. The cGMP level is in itself maintained by

CNP which stimulated cGMP production in the granulosa and cumulus cells. Thus oocyte maturation is associated with a fall in follicular cGMP and an associated LH surge which together lead to activation of oocyte-specific phosphodiesterases and fall in oocyte cAMP. This causes activation of maturation promoting factor leading to meiotic resumption.

3. Estradiol

The pre-ovulatory follicle is characterized by high levels of estradiol production and secretion. The primary role of building levels of estradiol is initiation of LH surge which leads to re-initiation of oocyte nuclear maturation and morphological restructuring of the follicle converting the dominat follicle into pre-ovulatory follicle, capable of releasing the oocyte for fertilization. Estradiol also leads to formation of new hyaluronan-rich matrix by the cumulus cells, initiated by the upregulation of matrix forming genes like HAS2 (Hyaluronan acid synthase), TSG6 (TNF-stimulated gene 6) and TSG14/*PTX3* (Pentraxin-related protein). This leads to spectacular increase in volume of cumulus cell layers that harbor the oocyte. This process is critical for normal fertility, as without it, the oocyte cannot escape the follicle and reach the oviduct to be available for fertilization.

4. Maturation promoting factor (MPF)

MPF activity drives somatic cells into mitosis and oocytes into meiosis. It was identified in progesterone- treated frog oocytes, as a factor capable of causing nuclear envelope breakdown by phosphorylation of condensin (protein complex needed to supercoil DNA during mitosis) and several protein components of nuclear envelope and components of nuclear pore complex like nucleoporin which are critical for nuclear envelope breakdown. It is a heterodimer composed of two subunits: i) Catalytic subunit, CDK1 (p34^{cdc2}); ii) Regulatory subunit, Cyclin B (B1, B2 and B3). Although the binding of CDK1 to Cyclin B1 is necessity for its kinase activity, the switching on of MPF is governed by a balance in regulatory activity of Wee1/Myt1 kinases and CDC25 phosphatases. These kinases cause an inhibitory phosphorylation of CDK1 at Threonine14 and Tyrosine15 residues which holds MPF in an active state, Pre-MPF. CDC25 phosphatases, on the other hand, cause an activating dephosphorylation of CDK1 at the same sites. Thus, high CDC25 and low Wee1/Myt1 activities are needed for switching on the CDK1 component of MPF. In mitotic cell, before entry into mitosis, Cyclin B1 (MPF) is restricted to cytoplasm only as it possesses a cytoplasmic retention sequence containing a nuclear export signal. Upon committing to mitosis, Cyclin B1 gets phosphorylated within its cytoplasmic retention sequence which leads to rapid accumulation of Cyclin B1 and MPF within the nucleus causing nuclear envelope breakdown. During meiotic arrest at the prophase I, MPF is phosphorylated by inhibitory proteins (Wee1/Myt1), keeping it in an inactive state. At the time of oocyte maturation, CDC25, a dual specific phosphatase, dephosphorylates CDK1 at the same residues, thereby activating it. All the three forms (A, B and C) of CDC25 are expressed in mouse oocyte but CDC25B is indispensable for resumption of meiosis. The active MPF then translocates from cytoplasm to nucleus leading to Germinal Vesicle breakdown (GVBD) by phosphorylation of nuclear proteins and nucleoporins. It has been observed that the concentrations of CDK1 and Cyclin B increase with oocyte growth and it is speculated that meiotic competence of oocytes may be dependent on a threshold of these proteins. It has been proposed that a balance of phosphorylated and unphosphorylated states of MPF as well as its spatial localization within the oocyte may underlie the event of oocyte competence. The kinase activity of Wee1/ Myt1 to maintain MPF in an inactive state is regulated by high levels of cAMP during the meiotic arrest. High levels of cAMP also inhibit the expression of Cyclin B, thereby decreasing the availability of pre-MPF. The elevated levels of cAMP during meiotic arrest activate protein kinase A which in turn regulates Wee1/Myt1 activity. PKA has two regulatory subunits (R1 and R2) whose expression varies with cAMP concentration and subcellular localization. PKA R1 subunit is highly sensitive to cAMP levels and is expressed predominantly in growing oocytes. PKA R2 is expressed at higher levels in fully-grown oocytes and is less sensitive to cAMP levels. PKA thus plays a crucial role in both meiotic arrest and maturation and is in turn regulated by a set of proteins called *A Kinase Anchoring Protein* (AKAP). During meiotic arrest, PKA binds to an isoform of AKAP in cytoplasm which results in trans-localization of PKA subunits to the site of action (near Wee1 and CDC25), thus causing meiotic arrest. During meiotic resumption and GVBD, PKA binds to another isoform of AKAP (AKAP1) and is relocalized to the mitochondria away from its site of action. Thus, AKAPs are involved in both meiotic arrest and release by regulating the spatial and temporal localization of PKA to its site of action. In mammals, the LH surge cause decrease in cAMP levels in an oocyte which activates MPF, thereby releasing oocyte from the meiotic arrest and enabling oocyte maturation. LH causes breakdown of gap-junctional complexes by phophorylating connexin 43 (*Cx43*) *via* the cAMP/PKA/ MAPK pathway. It also inhibits Cx43 translation, ultimately leading to elimination of gap-junctions. In bovines, Cx43 mRNA has been proposed as a marker of oocyte developmental competence, as its levels are significantly lower in poor quality COCs as compared to good quality COCs. Thus breakdown of cell-cell-oocyte communication appears to be one of the main factors involved in resumption of meiosis. Other signals induced by LH surge like EGF (Epidermal growth factor), AREG (Amphiregulin) and BTC (β-cellulin) also act on cumulus cells to cause cumulus expansion, gap junction breakdown and trigger an active signal which decreases cAMP levels, resulting in meiosis resumption. It has been shown that EGF-EGFR induces several genes like Cyclooxygenase 2 (Cox2), hyaluron synthase 2 (HAS2) and tumor necrosis factor stimulated gene 6 (TSG6) which play role in cumulus expansion and ovulation.

5. EGF-like peptides and EGF receptor

It has been demonstrated that EGF stimulates the expansion of cumulus cell matrix in isolated immature cumulus-oocyte complexes derived from antral follicles. Though EGF receptor (EGFR) was identified in granulosa cells but no evidence of EGF was found within follicular fluid, as the follicular mediator of LH-induced ovulatory signal. However, it was later discovered that LH surge induces a signal transduction pathway which causes rapid and short-lived expression of EGF family members like amphiregulin, epiregulin and beta-cellulin within both granulosa and cumulus cells. Activation of EGFR leads to signaling cascade involving extracellular signal-regulated kinases 1 and 2 (ERK1/2) which are recognized as the central elements in LH-induced signal transduction. The cascade leads to prostaglandin E2 (PGE2) production as well as up regulation of p38MAPK activity, which further promotes EGFR signaling in both granulosa and cumulus cells, amplifying the signal throughout the follicle.

6. Steroids

The role of gonadotrophin-induced steroids in oocyte maturation was known in amphibians and fish from early times. But it was perceived that steroids have no or little role in mammalian oocyte maturation. However, a number of studies have confirmed that the oocyte maturation process is conserved from amphibians to mammals and by demonstrating that androgens, estrogens and progestins trigger oocyte maturation both *in vitro* and *in vivo*. It has been demonstrated that EGF-EGFR in cumulus granulosa cells stimulates steroidogenesis by up regulation of steroidogenic acute regulatory protein (StAR) activity. The steroids produced trigger oocyte maturation *via* classical steroid receptors. Androgen treatment has also been shown to decrease cAMP levels, in addition to activating MAPK and CDK1 signaling pathways, thereby stimulating oocyte maturation. Porcine oocytes have shown to undergo germinal vesicle and gap junction breakdown in response to progesterone treatment. It has also been shown that Follicle stimulating hormone (FSH) induces the expression of androgen receptor (AR) and Cytochrome P450 lanosterol 14a-demethylase (CYP51) which is regarded as a key enzyme in biosynthesis of sterols and steroids that are involved in oocyte maturation.

Completion of meiosis I

After MPF activation, GVBD occurs and meiosis is resumed leading to meiotic spindle formation and extrusion of the first polar body. The completion of the first meiotic division is marked by segregation of the homologous chromosomes between the oocyte and first polar body, while sister chromatids remain attached. The oocyte continues meiosis II and gets arrested again at metaphase II. This progression of oocytes through meiosis I and transition to metaphase II stage involves the activity of different proteins, especially MPF. In rodents, pig and bovine, MPF acts in an oscillatory pattern increasing its activity at GVBD stage which declines at metaphase I completion and again increases at metaphase II, continuing to remain so until fertilization. The decline of MPF activity at metaphase I (MI) occurs due to proteosomal degradation of Cyclin B (1). The polyubiquitination of cyclin B1 for proteosomal degradation is caused by a multi-subunit E3 ligase complex, Anaphase-promoting comples/cyclosome (APC/C). APC/C is itself activated by CDC20 and

FZR1 which bind to specific sequences (D-box, KEN-box, CRY-box) on the target proteins and label them for degradation. Both these proteins are involved in maintaining low levels of Cyclin B1, thereby preventing MPF activation. During M-phase, activated MPF inhibits FZR1 by phophorylation to increase Cyclin B levels and promote further MPF activation. After meiotic resumption, as the oocyte progresses through MI, CDC20 accumulates while MPF phosphorylates and activates APC/C to bind CDC20 and form APC (CDC20) complex. This complex degrades Cyclin B and inactivates MPF resulting in progression of oocytes from MI to first meiotic anaphase. The timing of Cyclin B1 degradation is regulated by Spindle Assembly Checkpoint Proteins (SAC). They inhibit premature APC/C activation and thus prevent mis-segregation during MI. The main SAC proteins include MAD (Mitotic arrest deficient) and BUB (Budding inhibited by benzimidazole). The low expressions of MAD and BUB proteins have been proposed to be associated with oocyte ageing and are thought to be a factor causing aneuploidy in older women. Aurora kinase A (AURKA), a centrosome-localized serine/threonine kinase, is another protein involved in progression from MI to MII. It is expressed throughout the GV-stage in mouse oocyte but is activated only after GVBD and is localized to microtubule organizing centers. It is reported that AURKA may be involved in regulation of microtubule organizing centers, resumption of meiosis, spindle microtubule dynamics and organization of the metaphase spindle. The transition from meiosis I to metaphase II is brought about by MAPK pathway. This pathway is inactive in farm animals at GV stage but gets activated around GVBD and its activity steadily increases, reaching its maximum at MII. Inhibition of MAPK pathway at MI to MII transition has been shown to prevent chromosome condensation, first polar body extrusion and MII spindle formation. The activity of MAPK pathway is mediated by p90rsk and/or via Akt/PKB pathway. Phosphorylated MAPK and p90rsk are associated with microtubule assembly at different stages of oocyte maturation. At MII the oocyte is again arrested due to activity of cytostatic factor (CSF). The underlying mechanism involves ability of oocyte to maintain high MPF levels by inhibiting APC/C activity thus, preventing proteosomal degradation of Cyclin B. CSF can act by different pathways like: increasing MPF synthesis or through activating SAC proteins which inhibit APC/C activity or by activating Emi. Emi inhibits the ability of CDC20 to activate APC/C through mos-MAPK pathway. It also regulates proteosomal activity that prevents degradation of Cyclin B. Though a number of pathways seem to be involved, mos-MAPK pathway seems to be predominantly involved in MII arrest in pig and bovine oocytes. MAPK pathway activates MISS (MAPK-interacting and spindle stabilizing protein) and DOC1R that are involved in spindle stability during the MII arrest. MISS is stable only in MII-arrested oocytes, while expression of DOCR1 occurs at the time of meiotic maturation.

Oocyte ageing

The quality of oocyte deteriorates upon prolonged arrest at MII stage which significantly impacts fertility. This is termed as oocyte ageing. The aged oocytes have a higher tendency for spontaneous ovulation, decreased ability to fertilize and low developmental competence, high rate of DNA fragmentation and low MPF activity, as observed in farm animals. The aged oocytes have a high level of inactive pre-MPF that is proposed to be caused by an imbalance between kinase and phosphatase activities. Though exact mechanism of oocyte ageing is not known but a decrease in active MPF levels and degradation of cell cycle checkpoints are presumed to be the causative factors.

In vitro maturation of buffalo oocyte

There is an increasing interest in large scale production of embryos from buffalo, the principal dairy animal in the developing countries of Asia, especially India. The combined techniques of *in vitro* maturation, fertilization and culture (IVMFC) provide the excellent alternative for *in vitro* embryo production (IVEP) but IVMFC has shown limited success in terms of embryo yield, establishment of pregnancies and birth of calves, than anticipated. It has been understood that *in vitro* embryo development is strongly influenced by events occurring during oocyte maturation, fertilization and subsequent development of the fertilized oocytes. In this context, IVM assumes the primary importance for being the first event in the process and influencing the subsequent process of fertilization and embryonic yield. Since, oocytes are retrieved at an immature stage from all the follicles (2–8 mm diameter), and not only from the pre-ovulatory follicle, so all the oocytes collected are at an immature stage, demanding maturation under *in vitro* conditions for successful fertilization and embryonic development. This necessitates recapturing of LH surge and other physiological events associated with oocyte maturation in a culture dish. Hence, improvement in efficiency of IVM process is essential for successful IVEP. The important criterion is to mimic the *in vivo* process as exactly as possible and the attempts to simulate the physiological event under laboratory conditions has led to a large number of protocols, supplementation with growth factors, hormones, etc. for achieving efficient IVM of the oocyte. This has not only led to development of efficient culture regimes capable of supporting IVM, IVF and IVC to the blastocyst stage but has in the process generated huge variations in the process both within and between the laboratories. In this section a thorough discussion of various factors which have been shown to influence IVM is provided with the aim that researcher gains an insight of all the factors and identify an efficient protocol or devise his own protocol in the process. Considering that effective IVM is the foundation of embryo production, a number of factors on which this foundation relies are discussed below:

1. Oocyte yield

The recovery of large number of good quality oocytes remains an ultimate goal for the mass production of buffalo embryos. The number of oocytes as well as their quality in turn depends on:

a) Method of oocyte retrieval

Oocyte dissection results in significantly greater yields of highest quality oocytes than aspiration. Aspiration also leads to greater disruption of surrounding cumulus cells and does not always succeed in retrieving the highest-quality oocytes as their cumulus-oophorous is firmly attached to stratum granulosum which at times prevents oocyte retrieval by aspiration. Oocyte retrieval from live animals by ovum pick up (OPU) is also affected by numerous factors such as aspiration vacuum, hormonal pre-treatment of animals, puncture frequency, stage of estrous cycle and skill of the operator. For example, recovery rate of high quality oocytes decreases significantly with increase in vacuum pressure, while gonadotrophin pre- treatment increases oocyte recovery rates and blastocyst production. Higher number of competent oocytes was obtained when period between follicle stimulating hormone (FSH) administration and OPU was extended. It has been proposed that oocyte retrieval at growth phase of first follicular wave, before dominant follicle selection improves cattle OPU efficiency.

b. Temperature and time

The time interval between animal slaughter and oocyte recovery from the ovaries and the temperature at which the ovaries are transported to the laboratory play an important role in efficiency of IVM and blastocyst yield. The authors have experienced fair results when buffalo ovaries were transported in antibiotic fortified normal saline solution in thermos flasks to the laboratory within 3h of the slaughter. It has been suggested that exposure of COCs to temperature below 35°C during oocyte recovery significantly decreases both the quality and quantity of blastocysts produced.

2. Oocyte quality

Proper oocyte selection in the laboratory is crucial for successful embryo production. Presence of intact cumulus cell layers surrounding the oocyte and a homogenous cytoplasm are regarded as the best indicators of an immature oocyte to undergo maturation and embryonic development. Since cumulus cells provide nutrients to the oocyte during growth, play role in zona formation and synthesis of the matrix composed of proteins and hyaluronic acid, and are important in oviductal transport or in sperm trapping, so we developed a grading system of buffalo oocytes depending on the number of cumulus layers surrounding them. We classified buffalo oocytes aspirated from slaughtered ovaries as: **Grade 1:** compact cumulus-oocye complexes (COCs) with an unexpanded cumulus mass having ≥ 5 layers of cumulus cells and with homogenous cytoplasm; **Grade 2:** COCs with ≤ 4 layers of cumulus cells and with homogenous cytoplasm; **Grade 3:** oocytes without cumulus cells and with irRegular shrunken cytoplasm. When observed for maturation in IVM medium, composed of 10% fetal bovine serum (FBS) in TCM-199 with 5 µg/ mL porcine-FSH, we observed that Grade 1 COCs showed highest maturation and blastocyst yield compared to other grades. We showed that there is a clear relationship between oocyte morphology recorded after aspiration and the developmental potential of oocytes in terms of nuclear maturation, cleavage and morula/ blastocyst yield. Hence, *in vitro* embryo yield is lower in buffalo as compared to cattle as the availability of Grade 1 oocytes is only 0.4 per ovary in buffalo as compared to 12 per ovary in cattle. We suggest, based on our results, to also use Grade 2 oocytes, but not Grade 3, for IVEP in buffalo.

Table 1: Maturation rate, cleavage rate and subsequent development of different quality buffalo oocytes following standard IVMFC

Quality of oocyte	No. of oocytes that matured	No. of matured oocytes that cleaved	Proportion of cleaved embryos that developed to	
			Morula	Blastocyst
Grade 1	88/103 [A] (85%)	60/82 [A] (73%)	25/60 [A] (42%)	23/60 [A] (38%)
Grade 2	59/110 [B] (54%)	35/86 [B] (41%)	9/35 [B] (26%)	7/35 [B] (20%)
Grade 3	26/99 [C] (26%)	10/85 [C] (12%)	1/10 [C] (10%)	0/10 [C] (0%)

(Adapted from Chauhan et al. (1998). Reproduction fertility and development 10, 173–177). Different superscripts in the same column differ significantly ($p < 0.05$).

3. Media formulation

The culture medium affects both maturation rate as well as embryonic development. The media commonly used for IVM or IVC can be broadly divided into simple and complex media. Simple media are usually bicarbonate-buffered systems containing physiological saline with pyruvate, lactate and glucose. These media differ in ion concentrations and energy sources, thus introducing lab to lab variations. The complex media contain aminoacids, vitamins and purines in addition to basic components of simple media and are less heterogenous than simple media. Ours and most IVF laboratories routinely use TCM-199 as the basic IVM medium for cattle and buffalo. Other media like M-199, F-10, B2, RPMI-1640 and CR2 (Charles Rosenkran's medium) have also been reported. The choice of medium becomes important as it has been demonstrated that oocytes matured in medium leading to poor developmental competence have depressed levels of glycolysis than necessary for completion of maturation. The reduced level of glycolysis may reflect reduced activity of the pentose phosphate pathway, which plays an important role in meiotic maturation of bovine and buffalo oocytes. However, excessive glucose in the media impairs developmental competence, possibly due to an increase in reactive oxygen species (ROS) and decrease in intracellular glutathione content of cattle and bubaline oocytes. The addition of such agents, like β-Mercaptoethanol, which increase intracellular glutathione levels in cultured oocytes has been shown to improve both maturation rate and blastocyst yield, following IVC. IVM rates have been found to vary with the basal media used, like TCM-199, Ham's-F10, MEM, RPMI-1640, etc. Though there are contrasting reports about the medium resulting to highest IVM in buffalo oocytes, the maturation rates obtained using these media have been found acceptable. Most of the researcher's report highest maturation rate in TCM-199 medium.

4. Bovine serum type and other proteins

We demonstrated the efficacy of different types of sera on buffalo oocyte maturation. The maturation medium consisted of TCM-199 supplemented with 10% buffalo estrous serum (BOS), 10% superovulated buffalo serum (SBS), 10% steer serum (SS) and 10% fetal bovine serum (FBS). 10–15 Grade 1 oocytes were cultured in 50 µl droplets of the respective maturation media for 24 h in a CO_2 incubator at 38.5°C. We observed that the rate of cumulus expansion and nuclear maturation were not significantly different between the BOS, SBS, FBS and SS supplemented maturation media (Table 2). The blastocyst formation rate was however highest in SBS supplemented medium (Table 3).

Table 2: Effect of serum type on cumulus expansion and IVM of buffalo Grade1 oocytes

TCM-199 with	No. of oocytes expanded	No. of oocytes matured
10% BOS	75/224 (33%)	48/99 (54%)
10% SBS	70/219 (32%)	51/86 (59%)
10% FBS	110/286 (38%)	62/108 (57%)
10% SS	74/210 (35%)	39/68 (57%)

(Adapted from Chauhan et al.(1998). Reproduction fertility and development 10, 173-177). Different superscripts in the same column differ significantly (p<0.05).

Table 3: Effect of serum type on cleavage rate and subsequent development of buffalo embryos in IVM/ IVF/IVC

TCM-199 with	No. of oocytes inseminated	No. of oocytes that cleaved	Proportion of cleaved oocytes that developed to	
			Morulae	Blastocyst
10% BOS	185	104 [A] (56%)	18 [A] (17%)	15 [A] (14%)
10% SBS	135	87 [A] (64%)	15 [A] (17%)	28 [B] (32%)
10% FBS	164	59 [B] (36%)	33 [B] (63%)	12 [A] (23%)
10% SS	142	91 [A] (64%)	36 [C] (39%)	19 [A] (21%)

(Adapted from Chauhan et al.(1998). Reproduction fertility and development 10, 173-177). Different superscripts in the same column differ significantly ($p < 0.05$).

From the foregoing data we found that all the four types of sera are capable of supporting development of buffalo embryos up to blastocyst stage. The cumulus expansion and nuclear maturation rates were not significantly different among the sera types. In addition to our study, there are numerous reports on different types of sera employed for IVM and IVC. The reports are more often conflicting and the discrepancies could be attributed to differences in concentrations of various hormones, like estradiol, LH, FSH and TSH in different sera types from different laboratories.

In addition to sera types, various commercial products are available as serum substitutes for *in vitro* embryo production. For example Ultroser G at a concentration of 1–4% has been used successfully in cattle IVM without hormone supplementation. Polyvinylpyrrolidine (PVP) at a 0.3% concentration has also been employed as a serum substitute in absence of hormones for IVM and IVC of cattle embryos.

5. Hormones

Most of the IVM protocols employ luteinizing hormone (LH) or follicle stimulating hormone (FSH) or a combination of both for IVM. The effect of gonadotrophins and their relative importance on IVM of buffalo oocytes and subsequent fertilization and early embryonic development are still controversial. While some studies report no effect of LH on IVM, others claim significantly beneficial effects and higher maturation rates. A comparison between TCM-

199, CR1aa and mSOF (modified synthetic oviductal fluid) for IVM in buffalo oocytes showed that all the media resulted to higher maturation rate, when supplemented with LH (0.023 U/mL) than in its absence. It has been proposed that LH alters calcium distribution within the ooplasm and promotes increased glycolysis, increased mitochondrial glucose oxidation and increased glutamine metabolism within cumulus-cell enclosed oocytes. FSH has also been suggested to enhance IVM by enhancing cumulus cell expansion which in turn enhances sperm capacitation and the fertilization process. A significant increase in development of buffalo embryos up to blastocyst stage was found upon addition of FSH or eCG (equine chorionic gonadotropin) in IVM medium. FSH stimulates cAMP dependent protein kinase activity which plays a significant role in chromatin condensation and Maturation promoting factor activation, finally leading to meiotic resumption. Addition of growth hormone has also been reported to accelerate progression of meiosis and cumulus expansion. It also leads to enhanced cleavage rate and blastocyst yield. Growth hormone has been shown to influence oocyte maturation by affecting the kinetics of the first polar body extrusion. It also causes a better cytoplasmic maturation in terms of proper distribution of cell organelles or formation of male decondensation factor. TCM-199 + LH have been shown to cause greater buffalo oocyte maturation than TCM alone which in turn was higher than CR1aa +LH maturation medium. mSOF medium showed lesser maturation than LH and CR1aa both in presence and absence of LH. The presence of estradiol in IVM medium showed no effect on IVM but improved fertilization and early embryonic development in human oocytes. However, maturation of bovine oocytes in presence of high concentrations of estradiol had a negative effect on spindle formation and first polar body extrusion. Estradiol could however, be added at a concentration of 1μg/mL which almost equals to the concentration in follicular fluid of preovulatry bovine follicles shortly after the LH peak. The presence of estradiol in buffalo IVM medium has not shown any considerable effect on IVM rate.

6. Growth factors

The addition of growth factors like epidermal growth factor (EGF) stimulates the cumulus expansion and significantly increases the proportion of oocytes attaining maturation. Insulin-like growth factor-I (IGF-I), IGF-2 or transforming growth factor –α or β addition has been demonstrated to increase oocyte maturation rate. The maturation rate was, however, not affected when denuded oocytes were cultured with EGF, indicating mediation by cumulus cells. IGF-I has been proposed to be a major follicular factor responsible for stimulating oocyte maturation in buffalo. It has been shown to interact synergistically with FSH, with the later increasing IGF-I receptors in granulosa cells and hence interaction with FSH. FSH and IGF-I together act synergistically to enhance DNA synthesis, protein synthesis and steroidogenesis in presence of granulosa cells. The addition of LH, however, suppresses IGF-I and IGF-I + FSH stimulated oocyte maturation. When co-cultured with granulosa cells, addition of LH to cultures containing IGF-I and FSH caused a significant increase in oocyte maturation .

7. Follicular fluid

Follicular fluid (FF) is a serum transudate modified by follicular metabolic activities and contains specific constituents like steroids and glycoproteins synthesized by cells of the follicle wall. Supplementation of IVM medium by bovine follicular fluid (10–20%) has been shown to favour embryonic development in cattle. However, the high concentration (>30%) has been shown to suppress both resumption of meiosis, fertilization rate and embryonic development in bovines. We observed a significant increase in degree of cumulus expansion in presence of 20% and 40% buffalo follicular fluid, collected from 2–4 mm surface follicles of slaughter house derived ovaries, as compared to IVM medium containing 10% FBS and 10% FBS + 5 µg/mL porcine-FSH. A high degree of cumulus expansion with follicular fluid supplementation could be due to possible presence of gonadotropins and growth factors like IGF-I and TGF-α which are present in follicular fluid and have been reported to be the stimulators of cumulus expansion. The glycosaminoglycans present in follicular fluid have been reported to stimulate cumulus expansion in bovine oocytes without enhancing fertilizability of the oocyte. However, no difference in fertilization rate and proportion of embryos developing to morula or blastocyst stage was observed between follicular fluid and p-FSH containing IVM media. We further observed that simultaneous presence of FBS, p-FSH and follicular fluid did not result in any improvement in morula and blastocyst yields.

Table 4: Effect of supplementation of IVM medium with buffalo follicular fluid on nuclear maturation rates of buffalo oocytes in vitro

TCM-199 with	Maturation rate
10% FBS	47 a
10% FBS + p-FSH	74 b
20% buffalo follicular fluid	67 b
40% buffalo follicular fluid	67 b

(Adapted from Chuahan et al. 1997. Theriogenology, 48: 461–469).

8. Culture duration, type and temperature

The time required for IVM of mammalian oocytes varies with the species. In mouse it is 18 h, while goat oocytes need almost 27 h for efficient maturation. The maturation period for rabbit is 14–15 h, for cat it is 36 h and camel 24–48 h. Good quality buffalo and porcine oocytes require 22–24 h for IVM. Buffalo oocytes are incubated at 38.5°C in humidified CO_2 incubator (> 95% humidity) for efficient IVM. We observed a significant decrease in blastocyst yield and quality when buffalo oocytes were matured at 40.5°C and 41.5°C for the first 12 h as well as for complete 24 h of the IVM interval, as compared to control IVM conditions. The relative gene expression of heat shock proteins, pro-apoptotic genes and oxidative stress-related genes was significantly higher in oocytes matured at the higher temperature as compared to standard conditions.

Buffalo oocytes could be cultured singly or in groups of 10–20 oocytes for IVM. The two IVM systems (group and single) show significant differences in maturation rates. Group culture system (15–20 oocytes in 100 μL droplets of maturation medium) has been reported to yield higher proportion of matured oocytes than single oocyte culture system (1 oocyte in 5 μL droplets).

9. Oocyte co-culture

A number of contrasting reports are available on whether co-culture with granulosa, cumulus or oviductal cells increases or decreases buffalo oocyte maturation. In one of the studies in which buffalo oocytes were co-cultured with granulosa cells, cumulus cells and buffalo oviductal epithelial cells (3×10^6 cells/mL) highest maturation rate was observed in granulosa cell co-culture followed, respectively, by cumulus cells and buffalo oviductal epithelial cell co-cultures, as compared to the control IVM culture in which oocytes were put to maturation in absence of any somatic cell type. Co-culture of denuded oocyte with cumulus cells showed no effect on IVM rate and developmental competence, while the presence of intact COCs with denuded oocytes had a positive effect on IVM and their developmental competence. The presence of dispersed cumulus cells during IVM has also been found to improve developmental competence. This suggests that direct interaction between oocyte and cumulus cells is not essential during IVM and IVF of denuded oocyte. The diffusible factors, produced by cumulus cells and/or by the crosstalk between oocyte and cumulus cells in the intact complex, play a role in acquisition of developmental competence of the denuded oocyte.

Detection of oocyte maturation

Oocyte maturation can be judged directly by staining their nuclear and chromatin structure by nuclear stains (Hoechst, DAPI) and/or by the ability of the oocytes to be fertilized or activated. The ooplasmic changes (cytoplasmic maturation) that occur during maturation are still difficult to evaluate. These changes play a crucial role in assembling the correct metabolic environment for production of sufficient energy for cellular functions during maturation, cleavage and blastocyst formation. The degree of cumulus cell expansion can be used as a morphological indicator for maturation of oocytes (Figure 6.1). It has been observed that when fully grown oocytes are released from the follicles into the culture medium they resume meiosis spontaneously in maturation medium. However, the reduced development of *in vitro* derived embryos suggests that the conditions of IVM do not support cytoplasmic maturation. Thus, it becomes very important that the improvement of IVM system for oocytes are aimed at defining the *in vitro* conditions that are more similar to the *in vivo* environment so that IVEP efficiency gets a boost.

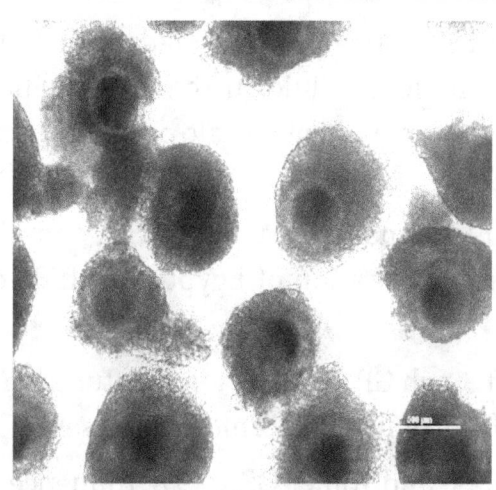

Fig. 6.1: Buffalo oocytes after IVM

Assessment of cumulus expansion

Cumulus cell expansion is assessed after 24 h of IVM. The matured COCs are examined under zoom stereomicroscope and degree of expansion (cumulus cell expansion score) is determined by visual assessment using the criteria described by Kobayashi *et al.* (1992):

Degree 0: No expansion;

Degree 1: Cumulus cells non-homogeneously spread and clustered cells still observed.

Degree 2: Cumulus cells homogeneously spread and clustered cells were no longer present.

Oocytes with moderate and fully expanded cumulus cell masses and unexpanded oocytes with an extruded first polar body in the perivitelline space are considered as mature.

Further reading

Kharche SD and Birade HS. (2013). Parthenogenesis and activation of mammalian oocytes for *in vitro* embryo production. A review. Advances in Bioscience and Biotechnology 4: 170–182.

Chauhan MS, Singla SK, Palta P, Manik RS and Madan ML. (1998). *In vitro* maturation, fertilization and subsequent development of buffalo (Bubalus bubalis) embryos: effects of oocyte quality and type of serum. Reproduction Fertility and Development, 10: 173–177.

Luciano AM, Lodde V, Beretta MS, Silvia C, Lauria A and Modina S. (2005). Developmental capability of fenuded bovine oocyte in co-culture system with intact cumulus-oocyte complexes: Role of cumulus cells, cyclic adenosine 3' 5'-monophosphate and glutathione. Molecular Reproduction and Development 71: 389–397.

Pawshe CH, Appa Rao KB and Totey SM. (1998). Effect of insulin-like growth factor I and its interaction with gonadotropins on *in vitro* maturation and embryonic development, cell proliferation and biosynthetic activity of cumulus-oocyte complexes and granulose cells in buffalo. Molecular Reproduction and Development, 49: 277–285.

Chauhan MS, Palta P, Das SK, Katiyar PK and Madan ML. (1997). Replacement of serum and hormone additives with follicular fluid in the IVM medium: effects on maturation, fertilization and subsequent development of buffalo oocytes *in vitro*. Theriogenology 48: 461–9.

Barakat IA, Kandeal SA, El-Ashmaoui HM, Barkawi A and El-Nahass E. (2012). Effect of mediu type and LH on *in vitro* maturation of Egyptian buffalo oocytes. African Journal of Biotechnology 11:4620–4630.

Totey SM, Pawshe CH and Singh GP. (1993). *In vitro* maturation and fertilization of buffalo oocytes (Bubalus bubalis): effects of media, hormones and sera. Theriogenology 39: 1153–71.

Das SK, Chauhan MS, Palta P and Tomer OS. (1997). Influence of cumulus cells on *in vitro* maturation of denuded buffalo (Bubalus bubalis) oocytes. Veterinary Record 141: 522–3.

Raza A, Samad HA, Rehman NU and Zia EU. (2001). Studies on *in vitro* maturation and fertilization of Nilli-Ravi buffalo follicular oocytes. International journal of Agriculture and Biology 3: 503–506.

Das GK, Jain GC, Solanki VS and Tripathi VN. (1996). Efficacy of various collection methods for oocyte retrieval in buffalo. Theriogenology, 46: 1403–11.

Ducibilla TP, Duffy R, Reindollar and Shu B. (1990). Cytoplasmic events of mammalian oocytes maturation. Biology of. Reproduction 43: 870–76.

Jainudeen MR, Takahashi Y, Nihayah M and Kanagawa H. (1993). *In vitro* maturation and fertilization of swamp buffalo (Bubalus bubalis) oocytes. Animal Reproduction Sciences, 31: 205–12.

Madan ML, Singla SK, Chauhan MS and Manik RS. (1994). *In vitro* production and transfer of embryos in buffales. Theriogenology, 41: 139–43.

Loose F, Van Vliet C, Van Maurik P and Kruip T. (1989). Morphology of immature bovine oocytes. Gamete research 24: 197–204.

El-Bawab IE. (1994). Meiotic maturation of Egyptian buffalo and cattle ovarian oocytes *in vitro*. Alex. Journal of Veterinary Sciences 10: 79–83.

Abdoon A, Kandil O, Otoi T and Suzuki T. (2001). Influence of oocyte quality, culture media and gonadotropins on cleavage rate and development of *in vitro* fertilized buffalo embryos. Animal Reproduction Science 65: 215–223.

Appendix 2

In vitro maturation of oocytes

We employ the following protocol for buffalo oocyte maturation in our laboratory.

The usable quality COCs are washed 4–6 times in the washing medium, followed by 2–3 washes in IVM medium (TCM-199 supplemented with 10% FBS, 5 mgml⁻¹ pFSH, 1 mgml⁻¹ estradiol-17β, 0.8 mM sodium pyruvate, 2.0 mM L-Glutamine and 50 mgml⁻¹ gentamicin). 15–20 COCs are then added into 100 μL droplets of IVM medium, overlaid with sterile mineral oil in 35 mm Petri dishes, and cultured for 24 h in a humidified CO_2 incubator (5% CO_2 in air) at 38.5°C. The matured oocytes were then used for IVF, parthenogenesis or Hand- guided cloning.

[Note: The Washing medium is the same medium as used for aspiration]

In vitro Maturation medium

Composition	Volume (10 ml)
Washing medium	10 ml
Porcine FSH	5 μg/ml
Follicular Fluid	500μl-1000μl
Estradiol 17–β	1μg/ml

Chapter 7

Sperm Capacitation – Preparation for the Union

Introduction

Spermatozoa are very peculiar cells in that they undergo several important maturation steps throughout their life in a coordinated manner and interaction with different environments where they develop, mature or travel through in order to accomplish the only function for which they are meant - fertilization. During maturation, spermatozoa get compartmentalized into acrosomal and post-acrosomal portions of head, mid-piece and distal flagellum with the aim of controlling the rate of modification differently in different compartments so as to efficiently co-ordinate its activity in an orderly manner. For example, in the epididymis, spermatozoa acquire the potential for motility and fertility by the action of epithelial cell secretions and constituents of the luminal fluids. The seminal fluid constituents at ejaculation bestow spermatozoa motility but the motile sperm is not yet able to fertilize the oocyte. It is during their migration through female genital tract that spermatozoa acquire the potential for which they are meant. It has been understood that spermatozoa are unable to fertilize oocytes unless they reside in the female genital tract for a specific period of time. During their migration through the genital tract, spermatozoa undergo a series of controlled biochemical and membranous changes that enable them to reach and bind to zona pellucida, undergo acrosome reaction, penetrate the egg membranes and finally fuse with the oocyte resulting in formation of an omnipotent cell - the zygote. This ensemble of transformations that spermatozoa must undergo to achieve fertilization capability is termed as capacitation.

Maturation of sperm in the testes

A sperm cell is a highly specialized cell acquiring a number of its characteristic features after the second meiotic division of the spermatocyte. After this division, the spermatids so formed are elongated to become sperm-shaped. This is followed by *Cap-phase*, in which acrosome is formed (form Golgi complex) that overlies the apical part of the elongated nucleus and contains a matrix of enzymes required by the sperm cells to reach the oocyte surface through cumulus layers and zona pellucida. During the *elongation period*, a characteristic condensation of chromatin takes place by replacement of histones with protamines. The formation of tail and its flagellum also takes place. Mitochondria are concentrated at the mid-piece, wrapped spirally around the central flagellum. Distal to the tail is the *Principal piece* which harbours a specific set of dense cytoskeletal structures around the central flagellum to bestow forward movement properties to the sperm. The most distal *End-piece* contains just the central flagellum surrounded by the plasma membrane. Upon liberation from the luminal surface of its guiding sertoli cell, the spermatozoa has shed off a number of organelles and cellular processes, like Golgi complexes, endoplasmic reticulum, lysosomes, peroxisomes and ribosomes, carrying with it only what is essentially required to deliver its haploid DNA content into the oocyte. With most of its cellular organelles shedded away, the sperm cell is neither capable of *de novo* protein synthesis nor of surface protein recycling due to absence of membrane vesicle transport, except at acrosome. Despite the molecular biological silencing of the sperm cell at spermiation, the sperm has already developed a polarized organization of intracellular organelles. The overlying plasma membrane interacts differently with different compartments resulting to different membrane subdomains that function independently and coordinately to accomplish the lengthy voyage through the male and female reproductive tracts till its final destiny in the ampulla of the female oviduct. Before capacitation in female reproductive tract, a sperm cell undergoes through two important processes of epididymal maturation and ejaculation.

The liberated sperm cell from the seminiferous tubule possesses a typical shape with a sperm head containing a nucleus with haploid genome, acrosome and tail. The sperm migrate through the retetestis into the epididymal duct at the caput epididymis to undergo epididymal maturation which involves final compaction of the protamine- DNA chromatin and removal of the cytoplasmic droplet. The cytoplasmic droplet is a remnant of cytoplasmic bridge that serves as syncytium of precursor germ cell stages and differentiating spermatids at the sertoli cell. After removal of this droplet, sperm cell virtually losses all its cytosol that was present in an early spermatid. In cauda epididymis, the sperm cell activates some metabolic pathways which provide it motility, unlike its immotile nature in caput epididymis and testis. The sperm cell also attracts a number of proteins at its surface which are known to be involved in fertilization. Some integral membrane proteins (glycosyl phosphatidyl inositol (GPI)-anchored) transfer form luminal plasma membrane of epididymal epithelial cells and bind covalently to sperm surface. This is followed by ejaculation where epididymal sperm cells are mixed with accessory fluid secreted by seminal vesicle glands, prostate, bulbouretheral glands, cowper glands etc. The presence of these glands and their relative volumes of secreted fluids vary among species. The mixing of these fluids causes sperm cells to encounter a completely new environment with a complex mixture of novel proteins. The seminal fluid stabilizes the sperm cells by coating their surface. These proteins are collectively called as decapacitation factors as they are believed to decrease the fertilization potential of ejaculated sperm. However, the binding of these proteins is very important for sperm survival as they enable the spermatozoa to survive the passage through vagina, cervix and uterus, the environments that are detrimental to sperm survival, by stabilization of the ejaculated sperm membrane. Seminal fluid also contains immunomodulaors that serve to protect the male genital tract and prevent immune responses towards sperm cells in the uterus. The extracellular vesicles (exosomes) in seminal fluid, like prostasomes and epididymosomes, also play a role in immune modulation. The presence of such exosomes, like uterosomes, has also been shown in uterus where they have been implicated in immune responses in female genital tract and in guiding the post-fertilization embryo that enters the uterus at a pre-implantation stage.

Sperm capacitation in female reproductive tract

At the moment of ejaculation, sperm exhibit a high level of progressive motility but are incapable of recognizing the egg or engaging in the complex cascade of cell-cell interactions that are essential for fertilizing the oocyte. During the ascent in the female reproductive system, spermatozoa successfully overcome the maternal immune system and ignore large number of other cells with which they make contact during their journey to the fallopian tube. On reaching the isthmic region of the oviduct, sperm behavior suddenly changes and the sperm establish intimate contact with the endosalpingeal epithelium, establishing a quiescent sperm reservoir and remain in this state until ovulation. At ovulation, the sperm suddenly break away from their epithelial resting place in a hyperactivated state and migrate rapidly towards the oocyte in a ready-to-fertilize state. By the time sperm reach the surface of oocyte, they are completely transformed cells exhibiting: a hyperactivated form of movement, various receptors for oocyte-cumulus mass on their surface and a plasma membrane primed to initiate the acrosome reaction in response to calcium transient. These modifications which a sperm undergoes are called as sperm capacitation and include multiple physiological and biochemical modifications like efflux of cholesterol from plasma membrane, increased membrane fluidity, increased permeability to bicarbonate and calcium ions, hyperpolarization of plasma membrane, changes in protein phosphorylation and protein kinase activity and increase in bicarbonate concentration, intracellular pH, calcium ions and cyclic adenosine monophosphate (cAMP) levels.

The whole process of capacitation can be divided into *fast* and *slow* signaling events which take place during the passage of sperm within the female reproductive tract. The fast events happen as soon as the sperm leave the epididymis and include activation of the vigorous and asymmetric movements of the flagella, while slow events include changes in movement pattern (hyperactivation) and protein tyrosine phosphorylation. The fast events are mediated by PKA activation and by Ca^{2+} and HCO_3^- - dependent soluble adenylyl cyclase. Ca^{2+} is transported into sperm by sperm-specific Ca^{2+} channel known as CatSper, while HCO_3^- is transported by Na^+/HCO_3^- co-transporters. The slow events are mediated by removal of cholesterol from the sperm membrane which increases sperm fluidity. All these changes together cause sperm capacitation bestowing to sperm the capability to: carry our acrosome reaction (AR), induced by agonists (zona pellucida and progesterone); produce hyperactivation motility (HAM); exhibit chemotactic behavior; and fertilize the oocyte to produce the omnipotent zygote.

Biochemistry of sperm capacitation

Tyrosine phosphorylation

Phosphorylation of proteins is a post translational event that acts as one of the cell's key regulatory mechanisms. Although both serine/threonine and tyrosine phosphorylation of proteins have been reported in spermatozoa, it is the tyrosine phosphorylation of a number of protein substrates that has been found to be associated with capacitiation in spermatozoa of most mammalian species. This phosphorylation is considered as the hall-mark event of capacitation. A number of tyrosine phosphoproteins have been identified in cattle and buffalo spermatozoa like serine/threonine- protein phosphate PP1γ2 catalytic subunit, MGC157332 protein, alpha-enolase, 3-oxoacid CoA transferase2 and actin-like protein 7A. In human, and probably, other mammalian species, tyrosine-phosphorylated proteins include ion channels, metabolic enzymes and structural proteins like CABYR, a calcium-binding protein localized in the principal piece of sperm tail. The main tyrosine-phosphorylated structural proteins of the fibrous sheath are the family of A-kinase-anchoring proteins which are involved in sperm motility. The primary kinases involved in triggering tyrosine-phosphorylation cascade are members of SRC-family particularly pp60cSRC and cABL. In addition to this, cAMP - mediated activation of protein kinase A (PKA) both directly activates these kinases and simultaneously suppresses an inhibitor of SRC, C-terminal SRC kinase. The targets of SRC-induced phosphorylation drive tyrosine phosphorylation *via* the phosphorylation-dependent inhibition of a tyrosine phosphatase, which normally keeps PKA-dependent tyrosine phosphorylation under inhibitory control. ERK (extracellular regulated kinases) pathway is also involved in capacitation of spermatozoa through receptor-activated tyrosine kinases like fibroblast growth factor, insulin-like growth factor receptor and epidermal growth factor receptor. The receptor kinases stimulate tyrosine phosphorylation by working through Ras-Raf-MEK-ERK network. The crosstalk has also been shown to occur between cAMP/PKA/SRC and MAP kinase pathways in regulating sperm tyrosine phosphorylation during capacitation. ERK pathway has also been shown to be activated by ROS, in absence of growth factor receptor activation, possibly as a consequence of phosphatase inactivation.

Protein Kinases A and C

It has been demonstrated that PKA is involved in regulation of sperm motility, as the activation of PKA catalytic subunit increases flagellar beat frequency during capacitation. PKA plays at least two independent roles in regulation of sperm motility- a fast action which is required for the activation of flagellar beat and a slow action which is required for change in flagellum waveform symmetry and requires PKA to be active for an extended period of time. It is accepted that during capacitation sperm change their motility from progressive to hyperactivated motility (HAM) which represents the capacitated sperm. PKJA has been undoubtedly shown to mediate HAM, a movement pattern characterized by asymmetrical flagellar beating observed in spermatozoa at the site and time of fertilization in mammals, and is presumed to be critical to fertilization process. It has been shown that hyperactivated sperm penetrate zona pellucida much more effectively than non-hyperactivated sperm. The prevention of HAM has also been shown to prevent fertilization.

PKC has also been shown to be involved in flagellar motility and acrosome reaction. It has been shown to be present in human, ram and bovine sperm and exists in a total of 11 isotypes, several of which can be simultaneously found in a single cell. The large number of PKC isotypes and the expression of most of them in sperm and eggs, suggests that this family of kinases has multiple tasks during gametogenesis, fertilization and early development. The PKC activators are produced by various phospholipases, and primarily include diacylglycerol (DAG) and phospholipids. They are either calcium-dependent or calcium-independent for their activity. PKC induces acrosome reaction in a calcium independent manner. The individual members of this kinase family are differentially regulated for fine tuning capacitation and other related events. The regulation is achieved by different cofactor requirement by different PKC isotypes, different substrate specificities of the individual family members and localization or enrichment of the specific isotype at specific locations in sperm or egg. For example, PKC is localized in the equatorial segment of the human sperm, while in bull sperm it is concentrated mainly in the post-acrosomal and upper region of acrosome.

P13K

P13K is responsible for production of phosphatidylinositol-3,4,5-triphosphate (PIP3), in response to growth factors, which is implicated in many biological processes like cell survival, cell growth, cell movement and adhesion, protein synthesis and cytoskeletal rearrangements. P13K catalytic and regulatory subunits are present in sperm suggesting its role in capacitation and acrosome reaction. A dual role of PKA has been found for regulation of P13K activity during bovine sperm capacitation. PKA first mediates P13K activation and then PKCα and PP1γ2 degradation. Both of these processes are necessary for P13 activation and production of PIP$_3$.

Actin polymerization and depolymerization

P13K activation by direct phosphorylation of p85 leads to actin polymerization. It is well known that actin polymerization occurs during sperm capacitation and that F-actin breakdown must take place to achieve acrosome reaction. An increase in F-actin is presumed to create a network between plasma and outer acrosomal membranes, while the dispersion of F-actin between the two membranes is needed to enable acrosome reaction. The presence of actin-binding proteins in mammalian sperm suggests that assembly of G-actin to form F-actin are well controlled events. Gelsolin severs assembled actin filaments and caps the fast growing plus end of free or newly severed filaments in response to calcium. Phosphoinositides bind gelsolin and release it from actin filament ends, exposing sites for actin assembly. It has been shown that gelsolin is inactive during capacitation and is activated prior to acrosome reaction. The release of gelsolin for phosphatidylinositol 4,5-bisphosphate, by PLC, causes rapid calcium-dependent F-actin depolymerization as well as an enhanced acrosome reaction. EGFR is also involved in acrosome reaction and in actin polymerization during capacitation. EGFR is partially activated in sperm incubated under capacitation conditions and is fully activated by adding EGF at the end of the capacitation, resulting in occurance of acrosome reaction. It has been suggested that during capacitation, intracellular calcium concentration rises which leads to conformational changes in gelsolin and exposure of F-actin binding site. The gelsolin is activated and transported to head of sperm. However, the concomitant elevation in PIP2 levels and gelsolin phosphorylation maintain gelsolin in an inactive state, allowing actin polymerization to occur. Prior to acrosome reaction, the intracellular calcium concentration is further elevated which activates PLC leading finally to hydrolysis of PIP2 and release of gelsolin to cytosol. The elevated levels of calcium and tyrosine dephosphorylation by tyrosine phophatases causes activation of the freed gelsolin which finally leads to F-actin dispersion and acrosome reaction.

Cholesterol and changes in membrane fluidity

The loss of cholesterol from sperm plasma membrane was one of the first changes described in capacitating mammalian spermatozoa. Cholesterol is a powerful decapacitation factor and serves to stabilize the plasma membrane of the spermatozoa during epididymal transit and prevents the intermolecular interactions responsible for achieving a capacitated state. Sterols like cholesterol and desmosterol are removed from the sperm surface by proteinaceous acceptor molecules such as albumin, high density lipoproteins and apolipoproteins, present in the extracellular space. A small percentage (~6%) of sperm cholesterol is stabilized in the sperm plasma membrane as cholesterol sulfate. As sperm ascend the female reproductive tract and initiate capacitation, sterol sulfatases affect the enzymatic hydrolysis of the sulfate group, thereby increasing the pool of cholesterol available for esterification. Phospholipase A provides the fatty acids required for esterifcation of cholesterol by enzymatic cleavage from membrane phospholipids. This generates highly unstable lysophospholipids which increase membrane fluidity and permeability to calcium ions, both of which promote capacitation and acrosome reaction. An active cholesterol transporter (ABCA17) is presumed to be involved in cholesterol transfer from sperm plasma membrane to albumin but oxidative stress is thought to be the major contributor. It has been shown that sterols become oxidized during capacitation which increases their hydrophilicity and facilitates their transfer to albumin.

Redox regulation

The involvement of reactive oxygen species (ROS) like hydrogen peroxide, superoxide anion and peroxynitrite anion in capacitation of mammalian spermatozoa has been clearly appreciated. These oxidants lead to suppression of tyrosine phosphatase activity and subsequent elevation of tyrosine phosphorylation levels that accompany capacitation. Superoxide also participates in direct activation of soluble adenylate cyclase thereby increasing the intracellular levels of cyclic adenosine monophosphate (cAMP) that, in turn, drive tyrosine kinase activity *via* a

Src-dependent mechanism. Hydrogen peroxide also enhances adenylyl cyclase activity by induction of tyrosine kinase activity which creates a self perpetuating cascade involving ROS generation, adenylyl cyclase activation and tyrosine phosphorylation. Peroxynitrite inhibits tyrosine phosphatases and activates tyrosine kinases of the Src family, making it a powerful contributor to the capacitation process. The scavenging of ROS by addition of catalase has been shown to prevent tyrosine phosphorylation surge associated with capacitation in human, hamster, buffalo, mouse and equine spermatozoa. The generation and involvement of ROS in capacitation places sperm on a knife-edge because sperm are inherently vulnerable to oxidative stress. It is a known fact that sperm capacitation and the entry into intrinsic apoptotic cascade is a continuum. It is for this reason that vitamin E has been shown to help preserve the functional integrity of spermatozoa by virtue of their capacity to counteract the oxidative stress associated with apoptotic death.

Hyperactivation of the capacitated sperm

Hyperactivation is a type of motility spermatozoa display at the site of fertilization. It is described as a non-synchronous, vigorous, whiplash type, frantic, high amplitude sperm motion. It refers to transit in the flagellar wave from the low-amplitude, symmetrical beat pattern typical of progressively motile sperm to a high amplitude, symmetrical thrashing of the sperm tail. Although the motility patterns and temporal relationship with capacitation may vary from one species to other, sperm hyperactivation appears to be an essential event of capacitation. Hyperactivated spermatozoa display a typical high velocity figure-of-eight pattern of movement that is thought to generate the propulsive forces necessary to pull the spermatozoa away from the oviductal epithelium and penetrate the dense matrix of zona pellucida. In some species, such as hamster, there is an orderly, relatively synchronized, progression towards a hyperactivated form of movement as the spermatozoa attain the capacitated state. The human spermatozoa exhibit brief transient unsynchronized bursts of hyperactivated movement as they become capacitated. A star-spin type of sperm motility, characterized mainly by very low linearity, is observed during sperm hyperactivation and appears to be a good predictor for the success of human *in vitro* fertilization. This occurs due to cAMP mediated tyrosine phosphorylation in sperm tail. This facilitates pulsatile pattern of calcium release in the flagellum from an intracellular calcium store thought to reside in the redundant nuclear envelope located at the base of the sperm head. Hyperactivation is believed to be due to changes in concentration of ions, HCO_3^- and Ca^{2+}, and inert-dependent changes at the cytosolic and axonemal levels. Potential axonemal modifications are related to phosphorylation of proteins known to be involved in sperm motility. Two proteins, p105 and p81, located on sperm fibrous sheath have been found to be progressively phosphorylated on tyrosine residues when human spermatozoa were incubated in capacitating conditions. The level of phosphorylation significantly increased after 1 h of incubation and further increases for up to 3 h of incubation, an interval of time that corresponds to the acquisition of hyperactivated motility in human spermatozoa incubated *in vitro*. Despite being an event of primary importance it is worth to mention that hyperactivation is not a synchronous process in the sense that not all the spermatozoa from a preparation display hyperactivated motility at the same time of incubation. Spermatozoa constantly switch from one pattern of hyperactivated motility to another or from non-hyperactivated to hyperactivated motility, making observations difficult.

Expression of oocyte receptors

One of the most dynamic properties acquired by capacitated spermatozoa is an ability to recognize zona pellucida. Earlier a number of sperm surface receptors were proposed to mediate this process including zona receptor kinase, mannosidase, sperm protein (SP) 56 and beta-glactosidase. Upon observation that the knoc-out mice lacking each of these putative receptors were fully fertile, an alternate mechanism was proposed which poised that a number of receptors are assembled into multimeric recognition complexes under the influence of molecular chaperones that finally lead to zona pellucida binding. For example, in mouse spermatozoa, the chaperones HSP90B1 (endoplasmin), HSPD1 (HSP60), as well as t-complex chaperonin family become surface oriented during capacitation in a phosphorylation-dependent process. They also reside within lipid rafts, microdomains that are moved within the plasma membrane in order to become localized at the anterior acrosomal aspect of the sperm head during capacitation. HSPA2 has been found to be chaperone facilitating human sperm binding to zona pellucida. The other two proteins that have been implicated in sperm-egg interactions are sperm adhesion molecule 1 (SPAM 1) and arylsulfatase A (ARSA).

In vitro sperm capacitation

Mammalian sperm are not immediately capable of fertilizing oocytes, rather they must undergo a period of preparation that normally occurs in female reproductive tract. The changes that occur in sperm involve at least two components: i) Initial sperm membrane alteration; and ii) fusion of the plasma membrane and an underlying outer acrosomal membrane so as to release acrosomal contents to enable fertilization. The first phase is considered to be the period of capacitation, and the second phase is the acrosomal reaction (AR). Two approaches have been used in a number of species to identify and differentiate the factors that capacitate sperm and factors that directly induce the AR. The first involves incubation of sperm under capacitating conditions, followed by exposure of sperm to oocytes for short periods of time. The central idea is that if the sperm are capacitated after the initial incubation, they will then fertilize oocytes within the time constraints imposed otherwise they will fail to fertilize the oocyte. The second approach is based on exposure of sperm (15–20 min) to substances that will induce an AR only in capacitated sperm. Ideally this substance would be an *in vivo* AR-inducing agent which may be the zona pellucida or other components of the cumulus-oocyte complex. As the *in vivo* AR-inducing substance is unknown for most species, a number of other factors like fusogeneic lipids and Ca^{2+} ionophores have been used for the purpose. AR has also been found to be induced *in vitro* by progesterone, albumin, phorbol esters, heparin, caffeine and by external manipulation of Ca^{2+} concentration. It has been demonstrated that capacitaion is an essential pre-requisite for acrosomal reaction and fertilization of the oocyte.

In vitro capacitation medium

Various media formulations have been prepared to achieve capacitation *in vitro*. The composition of these media varies both from species to species as well as within the species but in most cases the capacitating medium contains proper ions, energy substrates and albumin. There is apparently contradictory information in the literature on the contribution of each of the medium components to capacitation process. It is being postulated that there is no component whose presence in the medium is absolutely necessary for capacitation. This has been demonstrated for K^+, Ca^{2+}, HCO_3^-, albumin and others. A number of studies have also reported that *in vitro* capacitation is not absolutely essential for inducing the AR. However, all these ions perform important roles for the capacitation process which are described in the following section.

Role of bicarbonate ions in capacitation medium

The bicarbonate ions (HCO_3^-) enter into sperm through the co-transporter Na^+ / HCO_3^-. The physiological levels of HCO_3^- produce a rapid collapse of the asymmetry of the sperm plasma membrane. This is attributed to activation of enzymes that translocate membrane phospholipids, phosphatidylserine and phosphatidylethanolamine, and increase the availability of cholesterol to external acceptors. The increase in HCO_3^- concentration also produces an increase in intracellular pH and the activation of a unique type of adenylyl cyclase present in sperm. This soluble adenylyl cyclase results in increase in levels of cAMP and cAMP-dependent PKA activation. This modulates the response of calcium channels such as CatSper which produce changes in membrane potential and increase in intracellular Ca^{2+} concentration. PKA also phosphorylates several proteins on Ser and Thr residues, activating either directly or indirectly several protein kinases and/or inhibiting protein phosphatases which finally leads to increase in phosphorylation of Tyr residues. In sperm exposed to HCO_3^-, cAMP rises to its maximum levels within 60s, and the increase in PKA-dependent phosphorylation begins within 90s. This increase in tyrosine phosphorylation is a late event which depends on the presence of albumin, Ca^{2+} and HCO_3^- in the capacitation medium. The level of Tyr phosphorylation in human sperm has been shown to correlate strongly with sperm-zona-binding capacity. Alterations in Tyr phosphorylation have been found in subfertile subjects, indicating its physiological role in fertilization.

Role of calcium ions in capacitation medium

AR requires free extracellular calcium ions. A calcium-dependent isotype of PLCγ is activated during sperm capacitation, leading to activation of a broad range of PKC isotypes at the time of capacitation. PKA activates a voltage-dependent Ca^{2+} channel in the outer acrosomal membrane that releases Ca^{2+} from the interior of the acrosome to the cytosol. This rise in Ca^{2+}

levels further activates PLC, followed by formation of IP$_3$. This further elevates Ca^{2+} levels by mobilizing an acrosomal Ca^{2+} pool and the formation of DAG which activates specific PKC isoforms. The sperm plasma membrane contains a Ca^{2+} channel that is activated by PKC. PKC activation leads to increased sperm motility and AR. Ca^{2+} ions, *via* PLC activation, are also required for rapid F-actin depolymerization as well as enhanced AR.

Role of EGF in capacitation medium

Epidermal growth factor receptor (EGFR) signaling has been shown to be involved in actin polymerization during capacitation as well as in AR. It has been shown that EGFR is partially activated in sperm incubated under capacitation conditions but is fully activated by adding EGF at the end of capacitation resulting in occurance of AR. A7-nicotinic-acetyl-choline-receptor (α7nAChR) is presumed to be a potential sperm receptor that is activated by the egg zona pellucida to induce EGFR-mediated AR. In boar sperm and probably in other species, EGFR is involved in sperm motility. It is localized to higher extent in the acrosome region than in post-acrosome and flagellum regions, pointing to its role in AR. Proteomic analysis has also identified EGF-signaling as an important pathway in high fertility sperm.

Role of heparin in capacitation medium

Glycosaminoglycans (GAGs) are essential components of the extracellular matrix, contributing to cell recognition, cellular adhesion and growth regulation. Heparin or heparin sulphate is one of the four groups of GAGs. Heparin has been extensively used in *in vitro* studies to understand the endogenous role of heparin-like GAGs secreted by the epithelium of the female reproductive tract. The capacitating effect of heparin has been established in bovine, buffalo and human sperm. In cattle, heparin is thought to promote capacitation by binding to and removing seminal plasma proteins that are adsorbed to the sperm plasma membrane (spermadhesins) and would inhibit capacitation. It has also been shown to produce a rise of sperm intracellular pH and Ca^{2+} concentration, protein phosphorylation and modification of motility parameters. It has been observed that heparin binds to sperm in a specific and saturable manner and the binding sites are located mostly on the acrosomal region. In cattle and other species, heparin and other GAGs are powerful modulators of sperm binding to oviductal cells *in vitro* and are able to cause an increase of flagellar beating, followed by release of sperm with higher linear motility.

Role of caffeine in capacitation medium

As discussed, sperm capacitation reorganizes membrane proteins, alters metabolism of membrane phospholipids, causes reduction in membrane cholesterol levels and sperm hyperactivation to finally lead to AR. Caffeine has been found to play a role in all these events of sperm capacitation. Caffeine is an inhibitor of cyclic nucleotide phosphodiesterase, resulting to an increase in intracellular cAMP which stimulates capacitation and AR.

Role of albumin in capacitation medium

Serum albumin and energy substrates in capacitation medium function as a sink for removal of cholesterol from sperm plasma membrane which plays an important role in sperm capacitation and acrosome reaction, as already discussed. Thus albumin, together with high density lipoproteins and apolipoproteins, serves as proteinaceous acceptor molecule leading to removal of cholesterol. It is for this reason that albumin is an extremely important component of *in vitro* fertilization media.

Role of Vitamin E in capacitation medium

Mammalian spermatozoa membranes are rich in polyunsaturated fatty acids (PUFAs) that are sensitive to oxygen-induced damage mediated by lipid peroxidation. Thus sperm are sensitive to reactive oxygen species (ROS) attack which leads to reduced motility, decreased viability, axonemal damage and mid-piece morphological defects. The presence of antioxidants in capacitation medium would scavenge the ROS and consequently protect the sperm membrane. Vitamin E is a well known antioxidnant and has been shown to inhibit free radical damage to sensitive cell membranes. It acts as a scavanger of lipid peroxyl and alkoxyl radicals. Vitamin E also allows for physiological production of ROS that are essential for membrane changes required for capacitation, acrosome reaction and *in vitro* fertilization. The presence of ROS has been implicated to be one of the drivers for sperm capacitation. It places the sperm on knife-edge because of the fact that sperm are inherently vulnerable to oxidative stress. It has been demonstrated that sperm capacitation and the entry of these cells into the intrinsic apoptotic cascade are a continuum. The ROS that drive tyrosine phosphorylation, cAMP production and cholesterol efflux from the plasma membrane ultimately induce a state of apoptosis in the sperm, thereby zeroing all the good they have done. It is for this reason that antioxidants such as vitamin E have been shown to help preserve the functional integrity of spermatozoa by virtue of their capacity to counteract the oxidative stress associated with apoptotic death.

Role of Osteopontin in capacitation medium

Osteopontin (OSP), also known as phosphoprotein 1, is an acidic glycoprotein with a cell-binding domain comprising GRGDS (glycine-arginine-glycine-aspartate-serine) which is present in the female reproductive tract. It belongs to family of integrins and is synthesized in the bovine oviductal epithelium and released into oviductal fluid. Integrins, like OSP, are present on the surface of oocytes and spermatozoa and have been shown to facilitate the process of fertilization in mammals by promoting sperm-oocyte binding. OSP has been shown to enhance fertilization and subsequent embryo production by improving *in vitro* sperm capacitation in buffalo. The combination of OSP and heparin showed tremendous impact on sperm viability, suggesting a strong synergistic action.

Assessment of sperm capacitation

Although the exact molecular mechanisms of capacitation are not completely known, a range of studies have demonstrated involvement of numerous structural and biochemical modifications in spermatozoa such as changes in plasma membrane structure, cholesterol efflux, increase in membrane fluidity, increase in bicarbonate, calcium and cAMP levels, rise in pH, protein phosphorylation and protein kinase activity. All these changes are presumed to be necessary for initiation of sperm-egg binding and AR reaction. Protein tyrosine phosphorylation is one of the major events associated with capacitation, as observed in various mammalian species such as mouse, human, bovine and buffalo. It is known to regulate many sperm functions like motility, zona pellucida recognition and acquisition of fertilizing ability.

The ideal probe to assess sperm capacitation should not interfere with sperm physiology and should be easily available and specific for capacitated sperm, but such an ideal probe does not exist as of now. Most often, sperm capacitation is evaluated by changes in the expression and/or distribution of cell surface molecules, chlortetracycline staining and treatment of spermatozoa with substances known to induce the acrosome reaction preferentially in capacitated spermatozoa. As discussed already, sperm capacitation involves redistribution, modification, removal or appearance of glycoproteins, surface sugars and sugar binding proteins, as well as various other antigens. These modifications can be assessed by use of specific antibodies, lectins, sugars conjugated to albumin, etc. Some of the most commonly used methods for measurement of sperm capacitation are discussed in this section.

1. Chlortetracycline assay

The evaluation of capacitation and AR by changes in tetracycline fluorescence patterns was first performed with mouse spermatozoa and was subsequently applied to human, monkey, bull and buffalo spermatozoa evaluation. The principle of this assay is based on the changes in fluorescence when chlortetracycline chelates with membrane-associated divalent cations, mainly Ca^{2+}. This assay offers the advantage of measuring directly the percentages of non-capacitated, capacitated and acrosome-reacted spermatozoa in the same preparation. In this assay, fresh or frozen/ thawed buffalo sperm are treated with chlortetracycline (CTC) staining solution. CTC solution is composed of 750 mM chlortetracycline, 5 mM cysteine in 130 mM NaCl and 20 mM Tris HCl. It is prepared fresh and its pH is adjusted to 7.8 and stored at 4°C under dark conditions. Equal volumes of sperm suspension and CTC solution are mixed in a falcon tube at room temperature, to which 2μl of glutaraldehyde (12.5% in 20 mM Tris-HCl, pH 7.4) are added. 8 μl of sperm suspension from the falcon tube are placed on a clean slide, to which 2 μl of 0.22 M 1, 4-diaza-bicyclo (2,2,2) octane dissolved in glycerol: phosphate-buffered saline (9:1) are added to retard the fading of CTC fluorescence. The sperm suspension is covered with a cover slip and stored at 4°C overnight in dark. Chlortetracycline fluorescence is observed under the fluorescent microscope. The spermatozoa are classified into one of the three CTC staining patterns: 1) Uniform bright fluorescence over the whole head

(*Uncapacitated spermatozoa*); 2) Fluorescence-free band in post-acrosomal region (*Capacitated spermatozoa*); and 3) dull fluorescence over the whole head except for a thin punctuate band of fluorescence along the equatorial segment (*Acrosome reacted spermatozoa*).

2. Dual staining (Trypan blue/Giemsa) technique after LPC treatment

Lysophosphatidyl choline (LPC) is a membrane disturbing agent found at high concentrations in the region of the female genital tract where fertilization takes place. LPC was shown to induce acrosome reaction in capacitated guinea pig spermatozoa and was subsequently used to assess sperm capacitation across the species with varying concentration of LPC and albumin in the incubation medium to prevent potential toxic effects. In this method, capacitation is assessed indirectly by estimating the percentage of AR-sperm after 15 min incubation with optimized concentration of LPC (usually 60 µg/ml). LPC is a fusogenic agent known to induce AR only in capacitated sperm at 38.5°C temperature in a 5% CO_2 incubator. To further evaluate sperm viability and AR, sperm are fixed and stained with dual staining technique. Trypan blue is used first to differentiate live from dead spermatozoa. Then the dried smears are fixed in 37% formaldehyde for giemsa staining to evaluate acrosome reaction under a microscope. Based on staining characteristics, sperm are differentiated into 4 categories: 1) Acrosome-intact live (AIL); 2) Acrosome-intact dead (AID); 3) acrosome-reacted live (ARL); and 4) acrosome-lost dead (ALD). The live sperm displays both head and tail viable, while the dead are those which have either the head or tail unviable (trypan blue stained).

3. D-mannose binding lectins

The measurement of D-mannose binding lectins is physiologically relevant as these lectins are considered as putative zona-pellucida recognition molecules on sperm membrane. A strong correlation exists between the transportation of these lectins to the plasma membrane and success of *in vitro* fertilization.

4. Fibronectin expression

An increased expression or unmasking of cellular form of fibronectin on human sperm surface occurs during capacitation. It is presumed that fibronectin is involved in sperm-oocyte interactions since anti-fibronectin antibodies have been found to greatly reduce the adhesion and penetration of zona-free hamster eggs by human spermatozoa.

5. Immunolocalization of protein tyrosine phosphorylation

Since tyrosine phosphorylation is the major mechanism involved in sperm capacitation and acrosome reaction, measurement of the tyrosine phosphorylation will provide an accurate estimate of capacitation efficiency. The localization of phosphotyrosine containing protein is detected using an indirect immunofluorescence assay. Semen is fixed in 2% formaldehyde for 1 h at 4°C and pelleted at 300 g for 10 min. The sperm pellets are incubated overnight at 4°C in modified phosphate buffer saline (mPBS: 2.7 mM KCl, 1.5 mM KH2PO4, 8.1 mM Na2HPO4, 137 mM NaCl, 5.5 mM glucose and 1.0 mM pyruvate, *p*H 7.4) containing 2% BSA. After centrifugation (250 g, 10 min, 25°C), sperm pellets are resuspended and diluted 1:10 in mPBS. 20 µl of sperm suspension are smeared onto a slide, air dried and permeabilized with absolute ethanol for 5 min. The permeabilized spermatozoa are incubated with primary antibody against phosphotyrosines in tris buffered saline-TBS (20 mM Tris-HCl, 0.8% NaCl, *p*H 7.6) for 1 h at room temperature in a humid chamber. This is flowed by treatment with FITC-labelled secondary antibodies and fluorescence is observed under a fluorescent microscope. The capacitated spermatozoa show fluorescence in the equatorial segment or uniform fluorescence over the entire acrosome or fluorescence at both equatorial and anterior acrosomal regions, while the non-capacitated sperm do not show fluorescence. The better capacitated spermatozoa are those which show fluorescence at both the equatorial and anterior acrosomal regions.

6. Assessment by *in vitro* fertilization ability

It is a well established fact that proper capacitation is a pre-requisite for successful acrosome reaction and fertilization. The assessment of fertilization *in vitro* would therefore give an indication of sperm capacitation. For this purpose, oocytes are aspirated from buffalo ovaries and put for *in vitro* maturation, as already discussed. Buffalo frozen spermatozoa are processed in capacitation medium (Bracket-Oliphant medium) according to the protocol discussed in Appendix 3. The gametes are co-incubated for 18h in 5% CO_2 incubator at 38°C. The presumptive zygotes are washed off the cumulus cells and cultured in *in vitro* embryo culture medium for up to 48 h. After this period, cleavage rate is assessed and the percentage of cleaved embryos is recorded. Both the cleaved and uncleaved oocytes are dezonated by enzymatic digestion with 2 mg/ml Pronase for 3 to 5 min, fixed with absolute ethanol overnight and stained with 4, 6-diamidino-2-phenylindole (DAPI) for nuclear staining under a fluorescent microscope. The oocytes with two synchronous pronuclei (2 PN) are considered as normally fertilized (monospermic), as against more than 2 PN in polyspermic fertilization.

Despite a number of such methods being used for assessing *in vitro* capacitation, there is presently no ideal method for measuring the same. These methods should, therefore, be considered as tools to study sperm capacitation or compare fertile and infertile samples, while awaiting more efficient, reliable and ideal methodology. The ideal and physiological agonist to assess capacitation is the zona pellucida protein as it specifically binds to the capacitated spermatozoa and provokes acrosome reaction. Although solubilized zona pellucidae and purified zona pellucida (ZP3) have been used, these products are not easily available and are also very costly.

Semen cryopreservation and capacitation

Semen cryopreservation is considered as one of the most important tools in assisted reproduction. It has been observed that fertility of frozen/thawed semen is reduced by cryopreservation. A number of recent reports have recognized that cryopreservation procedures like dilution, cooling, freezing/thawing induce capacitation-like changes in spermatozoa. There are similarities in changes associated with capacitation and cryoinjury, such as plasma membrane reorganization and fluidization and calcium influx into the spermatozoa. Therefore, partially or fully cryopreserved spermatozoa demonstrate capacitation-like behavior and show the ability to undergo the acrosome reaction and fertilize oocytes *in vitro*. This so called, cryocapacitation, is considered to be partly responsible for the reduced longevity and fertility in artificial insemination of frozen/thawed bull semen.

Further reading

Parrish J, Susko P, Winer M and First N. (1998). Capacitation of bovine sperm by heparin, Biology of Reproduction, 38: 1171–1180.

Leclerc P, Sirard M, Ghafouleas J and Lambert R. (1992). Decrease in calmadoulin concentrations during heparin-induced capacitation in bovine spermatozoa. Journal of Reproduction and fertility, 94: 23–32.

Elkhawagah A, Longobardi V, Gasparrini B, Sosa G, Bifulco G, Abouelroos M, Abd el-ghafar A, Campanile G. (2013). Evaluation of *in vitro* capacitation of buffalo frozen/thawed sperm by different techniques. Journal of Reproduction and Infertility, 4: 19–28.

Oliveira V, Marques M, Simoes R, Assumpaco M and Visintin J. (2011). Influence of caffeine and chondroitin sulfate on swine sperm capacitation and *in vitro* embryo production. Acta Scientiae veterinariae, 39: 960.

Jaiswal B, Anat C, Tur-Kaspa I and Eisenbach M. (1998). Sperm capacitation is, after all, a prerequisite for both partial and complete acrosome reaction. FEBS letters, 427: 309–313.

Lamirande E, Leclerc P and Gagnon C. (1997). Capacitation as a regulatory event that primes spermatozoa for the acrosome reaction and fertilization. Molecular Human Reproduction, 3: 175–194.

Ickowicz D, Finkelstein M and Breitbart H. (2012). Mechanism of sperm capacitation and the acrosome reaction: role of protein kinases. Asian journal of Andrology, 14: 816–821.

Visconti P and Gregory K. (1998). Regulation of protein phosphorylation during sperm capacitation. Biology of Reproduction, 59: 1–6.

Salicioni A, Platt M, Wertheimer E, Arcelay E, Allaire A, Sosnik J and Visconti P. (2007). Signaling pathways involved in sperm capacitation. Spermatology, 65: 245–260.

Aitken R and Nixon B. (2013). Sperm capacitation: a distant landscape glimpsed but unexplored. Molecular Human Reproduction, 19: 785–793.

Appendix 3

The following protocol is being followed in our laboratory for *in vitro* capacitation of the cryopreserved buffalo sperm.

A straw of frozen thawed buffalo semen is washed twice with the washing Brackett and Oliphant (BO) medium (medium containing 10 mgml^{-1} heparin, 137.0 mgml^{-1} sodium pyruvate, 1.942 mgml^{-1} caffeine sodium benzoate and 50 mgml^{-1} gentamicin in DPBS). The pellet is resuspended in ~0.5 ml capacitation and fertilization BO medium (washing BO medium containing 10 mgml^{-1} fatty acid-free BSA). The *in vitro* matured oocytes are washed twice with the fertilization BO medium and transferred to 50 µL droplets (15–20 oocytes/droplet) of BO medium. For IVF, the spermatozoa in 50 µL of BO medium (2–4 million spermatozoa ml^{-1}) are added to the droplets containing the oocytes, covered with sterile mineral oil and placed in a CO_2 incubator (5% CO_2 in air) at 38.5°C for 16–18 h.

Brackett and Oliphant (BO) medium
Solution A (Stock)

Sodium chloride	4.3092 gm
Potassium chloride	0.1974 gm
Calcium chloride dihydrate	0.2171 gm

Dissolve the above in 500 ml of distilled water. Mix 0.1 ml of 0.5% phenol red for coloring and indication of pH of the solution. A yellowish color will appear after addition of phenol red. Add penicillin (50 I.U/ ml) and streptomycin (5 µg/ml) for preservation of longer periods.

Solution B (Stock)

Sodium bicarbonate	2.5873 gm

Dissolve it in 200 ml of distilled water. Mix 0.1 ml of 0.5% phenol red for indication of pH of the solution. Pink color will appear. Add penicillin/ streptomycin as given above.

Working BO medium

Composition	Volume (50 ml)
Solution A	38 ml
Solution B	12 ml
Heparin	10 µg/ml
Sodium pyruvate	0.0068 gm
Caffeine sodium benzoate	0.0971 gm

BO medium for capacitation and fertilization

Working BO media	10 ml
BSA (Fatty acid free)	gm

Chapter 8
In Vitro Fertilization

Introduction

Once the ejaculated semen is deposited into female reproductive tract, spermatozoa undergo capacitation attaining capability to attach to the egg surface and penetrate its membranes resulting into deposition of its nuclear material into the ooplam. This process is referred to as fertilization. It is more a chain of events than a single isolated phenomenon, and interruption at any step in the chain almost certainly results in fertilization failure. The successful fertilization requires not only fusion between a sperm and an egg, but demands a range of changes that a sperm as well as an oocyte has to undergo before the fusion leads to successful fertilization. Successful fertilization also requires that only one sperm fuses with an egg.

Fertilization encompasses at least five steps that take place in a compulsory order. Recent evidence suggests that mammalian sperm is drawn to the egg by a chemo-attractant emitted by follicle cells surrounding the egg, rather than simply by a chance encounter. The chemo-attractants are presumed to be heat-stable peptides, and the whole process is dubbed as *sperm chemotaxis*. The spermatozoa, with an intact acrosome, then binds in a species-specific manner to zona pellucida of the egg. This is followed by acrosome reaction in the spermatozoa and then penetration of the zona coat. Having reached the perivitelline space between the egg zona pellucida (ZP) and plasma membrane, sperm binds to the plasma membrane and fuses with it. At this point the egg is said to be fertilized and is called a zygote. The free-swimming sperm are no longer able to bind to the ZP. At least five steps have been identified to be involved in fertilization in mice and probably all the mammalian species which are a major subject of this chapter.

Sperm binding to egg

The binding of the sperm to an egg zona pellucida is species-specific. It has been observed that egg and sperm from different species fail to bind *in vitro* but the restriction has been overcome by removing the zona pellucida coat (with proteases or low pH buffers), thereby allowing sperm to bind directly to the egg plasma membrane. This indicates that zona pellucida may posssess receptors that are recognized by sperm from the same species. These receptors are known as sperm receptors. The sperm also possess proteins, known as egg-binding proteins which are compatible with eggs from the same species. In mouse three zona pellucida glycoproteins (ZP1, ZP2 and ZP3) are identified and ZP3 functions as the main sperm receptor. It has been proved that acrosome-intact sperm recognize and bind to specific O-linked oligosaccharides located on serine residues (Serine 332 and 334) near the carboxy terminus of ZP3 polypeptide. Thus, like many other examples of cellular adhesion, binding of sperm to egg ZP is a carbohydrate-mediated event. At least twenty different sperm proteins have been implicated in species-specific binding of sperm to eggs. These include a variety of enzymes such as β-galactosyltransferase, α-fucosyl transferase, protein tyrosine kinase (ZRK) and phospholipase A2; lectin like proteins such as mannose- and galactose-binding proteins and spermadhesins; sperm proteins like zonadhesin and sperm protein 56 (SP56); as well as ADAM family members like sperm β-fertilin and cyritestin. It is generally held that for a given species more than one kind of egg-binding protein (EBP) is involved in binding of sperm to egg. Also sperm from different mammalian species use different EBPs during fertilization. Thus, a high degree of pathway specificity is achieved through a sequence of steps, each having moderate selectivity.

Acrosome reaction

Binding of sperm to zona pellucida is the easier part of fertilization which is followed by the daunting task of penetrating the zona pellucida. The sperm has developed an acrosome for the purpose, a huge modified lysosome that is packed with zona-digesting enzymes and located at the anterior part of sperm head. Shortly after binding to the egg ZP, sperm undergo cellular exocytosis, commonly known as acrosome reaction (AR). The acrosome is a relatively large Golgi-derived, lysosome-like organelle that overlies the nucleus in the apical region of the sperm head. Although the acrosome is surrounded by a continuous membrane, it is usually described as consisting of an inner and outer membrane. The inner membrane overlies the nucleus while the outer underlies the plasma membrane. AR involves multiple fusions between outer acrosomal membrane and plasma membrane at the anterior region of the sperm head, extensive formation of hybrid membrane vesicles and exposure of inner acrosomal membrane and acrosomal contents. Only sperm that have completed the AR can penetrate the ZP and fuse with egg plasma membrane. It is now accepted that ZP3 is the natural agonist that initiates the AR upon sperm binding to egg. The plasma membrane overlying the sperm head binds to thousands of copies of ZP3 which is sufficient to induce AR. The reaction is mediated by several signal-transducing components like G proteins, inositol-3,4, 5-triphosphate (IP_3) and its receptors, phospholipase C, Ca^{2+} and voltage-sensitive Ca^{2+} channels. ZP3 stimulation of sperm activates G proteins (G_{i1}, G_{i2} and $G_{q/11}$) in sperm, polarizes sperm plasma membrane (from ~ -60mV to ~ -30mV), activates Ca^{2+} channels (T type), increases *p*H (by ~0.3 units) and intracellular Ca^{2+} concentration (from ~ 150nM to ~ 400nM). It is believed that Rab3A GTPase and SNARES, the components that are essential for intracellular membrane fusion of somatic cells, are also present in mammalian sperm and may participate in membrane fusion during acrosome reaction.

Sperm penetration into egg

Acrosome-reacted sperm remain bound to ZP coat, apparently by binding to ZP2 in case of mouse and probably other mammals. The sperm must now penetrate ZP to reach and fuse with the egg plasma membrane. The penetration is achieved by combination of sperm motility and enzymatic hydrolysis, the latter being catalysed by an acrosomal serine protease called Acrosin. It is suggested that sperm motility is of overriding importance to zona penetration, allowing the knife-shaped mammalian sperm to basically cut its way through the zona pellucida.

Sperm-oocyte fusion

Once spermatozoa penetrates the Zona pellucida, it binds to and fuses with the plasma membrane of the oocyte. Binding occurs at the posterior (post-acrosomal) region of the sperm head. This region is capable of fusion only after the acrosome reaction has taken place. The binding of sperm to egg plasma membrane is thought to be mediated by a member of the ADAM family of transmembrane proteins on sperm and integrin $\alpha6\beta1$ receptors on egg. Two ADAM proteins, fertilin and cyritestin, have been studied in detail in mouse and are thought to interact with integrin in the egg plasma membrane through their disintegrin domains. Fertilin-β plays a role in binding of sperm to egg plasma membrane, while Fertilin-α has been implicated in subsequent fusion steps of sperm and egg. CD9 in the egg plasma membrane has also been found to play a vital role in sperm-egg fusion in mouse. It has been found that CD9 in the egg plasma membrane is intimately associated with integrin $\alpha6\beta1$ to which Fertilin-β binds. CD9, thereby, regulates the interaction between integrin and fertilin that is ultimately responsible for sperm-egg fusion. The specificity of the sperm to bind with egg, bypassing all the cells that come in its way is a remarkable process and has always fascinated the researchers to identify the mechanism involved. Recently Japanese researchers identified the Izumo protein which is displayed on the sperm surface. The named the protein Izumo after the name of Japanese marriage shrine. Its mate on the egg is identified as Izumo receptor or Juno. It is currently believed that Izumo-Juno pairing is the first known essential interaction for sperm-egg recognition in any organism.

Egg activation

Prior to fertilization, egg is in a quiescent state, arrested at metaphase of the second meiotic division. Upon sperm binding, it rapidly undergoes a number of metabolic and physical changes that collectively are called as *egg activation*. These events lead to cortical reaction, activation of egg metabolism, completion of meiosis II and initiation of mitotic cell cycle. There are two hypotheses regarding the activation of egg by sperm binding. According to *receptor* hypothesis, fertilizing sperm interacts with a specific egg surface receptor and this interaction leads to signal transduction and effector activation. It is currently believed that Izumo protein on sperm surface triggers Juno and activates the receptor. This in turn activates tyrosine kinases and subsequently phospholipase C. *Fusion* hypothesis postulates that following fusion of the sperm and egg plasma membranes, a soluble sperm-derived factor enters the egg's cytoplasm and activates pathways leading to egg activation. A novel PLC isoform, PLCζ, is presumed to be the equivalent of the sperm factor. The detailed mechanism of the activation process will be dealt somewhere else. The net effect is the rise of intracellular calcium which leads to meiotic reactivation and *cortical-reaction*.

Cortical reaction refers to a massive exocytosis of cortical granules which appear shortly after sperm-oocyte fusion. These granules contain a mixture of enzymes, including proteases, which diffuse into the zona pellucida following exocytosis from the egg. These proteases alter the structure of zona pellucida and induce *Zona reaction* which represents the block to polyspermy in most mammals. Zona reaction hardens zona pellucida so that the runner-up spermatozoa that have not finished traversing the zona pellucida by the time hardening occurs are stopped in their tracks. Zona reaction also destroys sperm receptors so that any sperm that have not bound to zona pellucida will no longer be able to bind it, let alone fertilize the egg. Following fusion of sperm and oocyte, the sperm head is incorporated into the egg cytoplasm. The nuclear envelope of the sperm disappears and the chromatin rapidly loosens from its tightly packed site in a process called *decondensation*. Other sperm components like mitochondria are also degraded in the process. Chromatin from both sperm and egg are soon encapsulated in a nuclear membrane, forming pronuclei. Each pronucleus contains a haploid genome. They migrate together, their membranes break down and the two genomes condense into chromosomes, thereby constituting a diploid cell – the Zygote.

In vitro fertilization

The establishment of culture conditions for successful union of sperm and egg under laboratory conditions simulating the female oviduct is referred to as *in vitro* fertilization (IVF). Successful IVF requires appropriate preparation of sperm as well as oocyte, and culture conditions that are favourable to the metabolic activity of the male and female gametes. The medium and culture conditions for IVF must be capable of providing the secondary oocyte and the capacitated sperm with conditions that permit sperm penetration to occur readily, leading finally to formation of the omnipotent zygote. IVF has been accomplished in more than 20 mammalian species. The first report of successful IVF was documented by Chang in 1959 for obtaining live rabbit offspring. He cultured freshly ovulated rabbit oocytes with *in vivo* capacitated spermatozoa which were obtained by flushing the uterine horns of females mated 12 h beforehand. The cleaved ova were transferred to oviducts of hormonally synchronous recipient does that delivered live offspring. It was recognized early in serious efforts to achieve IVF that the best results followed by rapid handling of gametes at near body temperature prior to insemination. Paraffin or silicone oil covering of media and high relative humidity assists in maintaining constant temperature by preventing evaporation. An oxygen tension of 8% to resemble that measured within the oviduct and 5% CO_2 to maintain proper *p*H of a bicarbonate-containing medium were found compatible with IVF. Thus, a simple defined medium consisting of salt solution with crystalline bovine albumin, glucose and bicarbonate to maintain pH of 7.8 to resemble estrous oviductal fluid at 38°C (rabbit body temperature) was found to consistently support sperm penetration of rabbit ova. This early work in rabbit provided a background for facilitating extension of IVF technology to other mammalian species. Also various factors were assessed for their influence on sperm capacitation and proportion of ova that could be fertilized. The cleavage of resulting zygotes was consistently obtained following transfer into a serum-containing medium with a more neutral pH. From embryo culture experiments, it became clear that Pyruvate is an important substrate that supports initial cleavage development. The need to transfer inseminated ova into another medium could also be eliminated by addition of 10^{-5} M Pyruvate to the defined medium, which already contains adequate concentrations of glucose and albumin to support later cleavage stages of preimplantation rabbit embryos. The medium developed for rabbit IVF has been variously referred to as Defined Medium (DM), Brackett's Medium and Brackett-Oliphant (BO) medium. IVF as well as *in vitro* sperm capacitation was performed in hamster ova by Yanagimachi and Chang in 1963–64. Hamster sperm heads and male pronuclei with tails of fertilizing spermatozoa were found within ooplasm after incubation of ova with epididymal sperm in Tyrode's solution under paraffin oil. Yanagimachi described microscopic observation of sperm penetration to include absence of the outer acrosome membrane before spermatozoon began penetrating the zona pellucida. He observed that the penetration was at an angle to the zona surface and required 3–4 min, and within less than 2s the sperm head lay flat on the oolemma. Thereafter, without any further motility the spermatozoon sank into the vitellus. Embryonic development beyond 2-cell stage was not achieved in hamster IVF, despite achieving 75% or more cleavage. This led to development of Hamster embryo culture medium (HECM), a modified Tyrode's medium with polyvinylalcohol present to replace previously used protein supplements (Bovine serum albumin) and an atmosphere of 10% CO_2, 10% O_2 and 37.5°C temperature. The modified culture conditions resulted into development up to blastocyst stage. The deletion of glucose, phosphate and pyruvate, reduction of lactate and addition of glutamine were important changes in the modified medium. Almost 30 years after the first successful IVF in golden hamsters, Barnett and Bavister reported that IVF hamster embryos could develop in chemically defined, protein-free culture medium (HECM-3 with hypotaurine) into morulae and blastocysts, and produce normal offspring after transfer to recipients. Thus, the zygote culture medium was different from sperm penetration and IVF medium which were supplemented with BSA. It was in 1968 when Whittingham reported *in vitro* fertilization and cleavage of mouse oocytes with spermatozoa recovered from 1–2 h post-coital uterus. Miyamoto and Chang in 1974 were successful in obtaining live young after transfer of 2-cell embryo, developed after direct *in vitro* insemination with epididymal spermatozoa. The IVF medium consisted of modified Krebs-Ringer bicarbonate solution supplemented with glucose, sodium pyruvate, sodium lactate, BSA and antibiotics. The time required for sperm penetration was shortened by sperm pre-incubation for 5–6 h and inclusion of high potassium to sodium ratio (0.32) and 2mM db-cAMP for IVF. The medium developed by Biggers, Whitten and Whittingham (BWW medium) was found to be adequate for achieving guinea pig IVF by Yanagimachi. In 1972, heterologous IVF experiments were successfully performed when Yanagimachi showed that *in vitro* capacitated guinea pig epididymal spermatozoa could penetrate hamster ova freed from their zona pellucidae, although penetration through intact zona was not possible. Handa and Chang also reported that rat and mouse spermatozoa were capable of penetrating zona-free hamster ova. This zona-free hamster ovum model was later advocated for evaluation of capacitation, acrosome reaction and male pronuclear development in species whose ova are difficult to study directly for any reason; e.g., human and pig. Hybrid embryos were also reported and their developmental capacity studied. The bull hybrid embryo obtained by fertilization of bovine oocytes with rat sperm developed up to 8-cell stage embryos but were incapable of undergoing maternal to embryonic genomic transition.

Among domestic animals, greatest progress in developing IVF and complementing technology has resulted from emphasis on bovine reproduction. As with development of IVF procedures in laboratory animals, a major challenge involved development of sperm treatments for ovum penetration. The initial success was facilitated by models made available by rabbit *in vitro* capacitation and through studies of bull sperm interaction with zona-free hamster ova. In 1977, Iritani and Niwa reported penetration and activation of around 20% of follicular oocytes following their maturation for 20–24 h at 37°C. The sperm penetration, as observed 19–24 h after *in vitro* insemination, was favoured by ova recovered from follicles and/or oviducts near the time of induced ovulation as compared to those recovered from un-stimulated 2–5 mm follicles and cultured for 18–25 h before insemination. The first bovine offspring from IVF was a bull calf born in June, 1981 from the experiments performed by Brackett and his group. They retrieved *in vivo* matured ova from donors treated with prostaglandin F2α and PMSG. The freshly collected semen was incubated in DM after high ionic strength-treatment. A range of fertilization results were observed according to different bulls that provided semen. Spermatozoa from one bull fertilized 22 of the 35 tubal ova and the *in vitro* embryonic development proceeded up to 8-cell stage. The embryos were surgically transferred into synchronized recipients and live birth followed. Variability in obtaining suitable ova following hormonal treatments, difficulties inherent in directly working with animals and high costs led scientists to devote serious attention to maturing oocytes easily obtained from small ovarian follicles at slaughter. *In vitro* matured slaughter house oocytes represent valuable experimental material for developing improved treatments for capacitation and acrosome reaction. A range of such studies led to identification of proteoglycans and glycosaminoglycans (GAGs) as the capacitating agents present at the normal site of fertilization *in vivo*. Heparin treatment was subsequently found to be effective for sperm capacitation and this GAG is extensively used in bull sperm treatments to affect *in vitro* capacitation. Although heparin/ heparin sulfate is the group of GAGs implicated most likely to be responsible for physiologically effecting capacitation on estrual cows, available evidence implicates other classes of GAGs, bicarbonate, calcium and other factors for physiological capacitation of the spermatozoa. Hyaluronic acid (HA) has been suggested as a capacitating agent for bull sperm capacitation. Swim-up procedures through columns of HA-supplemented media have enabled selection of viable and fertile bull spermatozoa for use in IVF. It is believed that by traversing the HA-containing medium, sperm surface is subjected to high shearing forces that might remove decapacitation factors thereby contributing to the capacitation process. The positive influence was found on bovine embryo development to the blastocyst stage following IVM and IVF after supplementation of chemically defined culture media with HA (1 mg/ ml). A number of other approaches have been employed for successful capacitation of bull sperm for IVF like: synergistic combination of heparin and caffeine; brief treatment with calcium ionophore A23187 with or without caffeine in DM; percoll separation followed by incubation with hypotaurine or caffeine; and calcium-free Tyrodes at *p*H 7.6.

A part of bull-sperm treatments for IVF involved selection of the most progressively motile sperm cell populations for insemination. This is commonly done by allowing the cells to swim up into media over an interval of 30–60 min after washing the sperm cells by centrifugation. Another method to achieve this involved centrifugation of spermatozoa through a gradient of 45 and 90% Percoll. Comparable cleavage and blastocyst rates were also reported after IVF with frozen-thawed bull spermatozoa separated with glass wool and by swim-up techniques. Transmigration (TM), suggested by Rosenkranz and Holzmann, selects motile spermatozoa in a target chamber after swimming from a test chamber through a unique membrane (8 μm diameter pores) against a stream of capacitating medium (5 ml/h). Using this system, bull-ejaculated spermatozoa did not need any additives like heparin, hypotaurine and epinephrine to promote capacitation and to increase motility. National Dairy Research Institute, Karnal reported the birth of first IVF- buffalo calf named Pratham in 1990, following the protocols similar to cattle IVF. The blastocyst developed after IVF of slaughter house derived oocytes was transferred to recipient cows, leading to birth of the live calf. Live IVF piglets as well as lambs were subsequently reported in 1986. Handa reported birth of a goat kid resulting from IVF of *in vivo* matured oocytes with A23187-treated spermatozoa. Modified DM was found to be superior to TALP or modified H-M199 for caprine sperm capacitation and better results were obtained with IVM, than with oocytes harvested with hormonal treatment of does. By 1994 conditions were described for successful IVM, IVF and IVC (*in vitro* culture) of goat embryos, enabling immature oocytes to develop into morulae and to kids after embryo transfer. Oocyte-cumulus complexes are matured for 27 h in TCM-199 supplemented with 20% fetal bovine serum (FBS), 100 μg LH/mL, 0.5 μg FSH/mL, and 1 μg estradiol 17- β/mL at 38.5°C in a humidified 5% CO_2, 5% O_2, and 90% N_2 atmosphere. Freshly collected spermatozoa were washed and incubated for 5 h in mDM-containing 20% FBS, then treated with 7.35 mM calcium lactate in the presence of ova for 14 h, followed by IVC on a cumulus-cell monolayer in HEPES buffered TCM-199 with 10% FBS under paraffin oil. 30 μl of spent medium was replaced by an equal volume of fresh HM-199 every 24 h. In 1991, Palmer and his group reported the first foal derived by IVF. They fertilized the pre-ovulatory oocytes by calcium ionophore A23187- treated spermatozoa.

Factors affecting *in vitro* fertilization

Fertilization is a complex process which results in the union of two gametes to restore the somatic chromosome number and start development of a new individual from the totipotent zygote. Fertilization is considered as the most critical step of *in vitro* embryo production procedures. In buffalo, cleavage rates lower than other domestic species have been widely reported. In addition to various factors affecting oocyte quality and oocyte maturation, success or efficiency of IVF is dependent on a number of factors such as sperm viability and capability, the adequate *in vitro* environment for gamete survival, the time of insemination, duration of gamete co-incubation, presence of cumulus cells, and acquisition of oocyte developmental competence during cytoplasmic maturation.

1. Preparation of sperm for fertilization

Fertilization involves a sequence of events in which the hyperactivated and capacitated sperm binds to egg zona pellucida to fuse with the vitelline membrane and undergo syngamy. Thus, it is important to have highly motile sperm available for IVF. The factors affecting sperm motility are discussed in this section.

i) Bull effect

A significant variability exists among the bull sperm to become capacitated. The individual bull variability is related to age of animal, season and ejaculate sperm quality. Seminal plasma has been ascribed to be a source of variability due to variation in level of de-capacitation factors in the seminal plasma.

ii) State of semen

Both fresh and frozen semen have been employed for IVF in cattle and buffalo. It has been found that fresh semen requires longer capacitation period than frozen semen. However, it has been found that fresh semen gives better penetration rates than frozen thawed semen. Frozen-thawed bull semen has also been reported to deteriorate more rapidly than fresh semen. It has been reported that post-thaw motility is an important factor in determining the success of IVF in buffalo. The bulls with high post-thaw motility produce significantly higher fertilization, cleavage and embryo developmental rates.

iii) Method of semen separation

A number of procedures, like Percoll density gradient, Swim-up, Sephadex, Glass wool etc. have been employed for sperm separation. The recovery rate of motile spermatozoa was found to be higher for sperm separated by percoll rather than swim-up method. However, swim-up procedure resulted in more ova being penetrated. The ion-exchange filtration was also found to be superior to swim-up procedure in harvesting maximum number of motile spermatozoa from frozen-thawed buffalo semen.

2. Capacitation and acrosome reaction

Capacitation is a process of involving the sperm in a complex series of biochemical and physiological reactions. It involves the removal and alteration of components derived from the seminiferous tubules, epididymis, vas deferns and seminal plasma to permit exposure of receptor sites that allows sperm to interact specifically with oocyte receptors. The efficiency and success of sperm capacitation further depends on:

i) Fertilization medium

Treatment of semen with medium of high ionic strength like Brackett and Oliphant medium (osmolarity 360 to 390 mOsm) has been used widely for semen capacitation. We have successfully employed the same medium, albeit with some modifications, for buffalo semen capacitation in our studies. TALP-medium has also been used for capacitation of bull and buffalo semen. A significantly higher blastocyst yield has been reported in oocytes inseminated with spermatozoa prepared by swim-up in Fert-TALP supplemented with heparin, than by centrifugation in BO supplemented with 10mM caffeine-sodium benzoate.

ii) Use of Heparin and Caffeine

It is being postulated that capacitation of bull or buffalo semen by heparin probably reflects the *in vivo* mechanism. Haparin dosage and incubation period for semen capacitation are important factors affecting bovine IVF and subsequently embryo development. Heparin has been found to induce changes in the calmodulin (CaM)-binding properties of serum proteins and induces a reduction in Ca^{2+} concentration during capacitation. It has been found that capacitation of bovine serum with Heparin requires extracellular calcium. The maximum kinetics of heparin-induced capacitation occurs when extracellular calcium exceeds 10 μM. The changes in Ca^{2+} concentration trigger increase in cAMP, pH and tyrosine phosphorylation that are essential for sperm capacitation. Frozen-thawed spermatozoa prepared in Bracket and Oliphant (BO) medium and treated with 5 mM caffeine + 10 μg/ml heparin, showed a higher fertilization rate as compared to the preparation in HEPES-TALP to which 10 μg/ml Heparin was added. A synergistic effect was also reported between Heparin and Caffeine for capacitation of frozen-thawed bull semen. 2.5 mM Caffeine and 10 μg/ml Heparin were the optimum doses for induction of capacitaion and acrosome reaction in buffalo spermatozoa. Upon comparison of fertilization rate between different Heparin dosages (0, 10, and 100 μg/ml) and sperm concentrations ($1x10^6$, $2x10^6$, $3x10^6$ and $4x10^6$/ ml) of frozen thawed buffalo semen from 6 bulls, fertilization rate was found to be significantly higher at 10 μg/ml Heparin and $2x10^6$ sperm concentration than the other groups.

iii) Cumulus cells

It has been found that removal of cumulus cells using hyaluronidase enzyme during *in vitro* fertilization of buffalo oocytes resulted in significant decrease in cleavage and embryonic development rates. The probable reason could be the role which cumulus cells play in sperm capacitation and acrosome reaction.

iv) Follicular fluid

The presence of follicular fluid in capacitation medium has been found to have positive effect on buffalo sperm capacitation. Pre-incubation of frozen-thawed sperm in follicular fluid also resulted to higher sperm penetration rate. The presence of GAGs in follicular fluid has been attributed for the increased sperm capacitation.

v) Use of calcium ionophore (A23187)

Influx of extra cellular Ca^{2+} is regarded as a primary event in sperm capacitation *in vivo*. Under *in vitro* conditions calcium ionophore A23187 has been employed for this purpose. It has been shown that ionophore treatment results in hyperactivation and a functional acrosome reaction in bovine sperm, enabling them to penetrate the zona-free hamster eggs. Comparable results have been obtained with Heparin and A23187 treatment, the advantage of the latter being its simplicity of use. However, it has been reported recently that calcium ionophore could not induce the acrosome reaction in absence of bicarbonate. It has been suggested that ionophore synergizes the bicarbonate-mediated induction of acrosome reaction.

v) Effect of glucose

It is reported that glucose inhibits heparin for inducing sperm capacitation in cattle. However, if oocytes are inseminated in chemically defined medium, glucose is required for stimulating IVF.

3. Oocyte preparation

IVF rate also depends considerably on the oocytes to be fertilized. The healthy and good quality oocytes are better fertilized than poor grade oocytes. Fertilization rate is also affected by the presence or absence of cumulus cells. In an effort to make the surface of the *in vitro* matured oocyte more accessible to sperm in IVF, researchers have attempted to remove some or all of the cumulus cell layers, either by mechanical stripping in suitably sized micropipettes, use of enzymes such as hyaluronidase, chemical agents such as sodium citrate or by vortexing. Though there are controversial reports regarding affect of cumulus removal on IVF rate; with some studies reporting increase while others reporting a considerable decrease. A number of studies have reported that cumulus cells are necessary at the time of IVF to maximize the incidence of acrosome reaction. Though the sperm penetration rate was similar in denuded and cumulus enclosed bovine oocytes, a significant increase in polyspermy was found in the denuded oocytes. Removing of cumulus cells also results in considerable increase in the incidence of defective ZPs and ooplasm abnormalities.

In vitro fertilization culture system

The medium employed in IVF system must be capable of providing the secondary oocyte and the capacitated sperm with the conditions, which will permit sperm penetration to occur readily. Most of the researchers have mentioned IVF-TALP supplemented with heparin and caffeine as the basic medium for preparation of sperm droplet. We have successfully carried out IVF of buffalo oocytes with frozen-thawed semen in BO medium supplemented with heparin and caffeine. The capacitated spermatozoa are added to the droplets at a concentration of approximately $1–1.5 \times 10^6$ spermatozoa/ml and the standard conditions for co-culture of spermatozoa and oocytes are 16–22 h at 39°C in an atmosphere of 5% CO_2 in air. We obtained successful fertilization of buffalo oocytes in groups of 10–15 in 50 µl droplets of fertilization medium, overlaid with sterile mineral oil, after insemination with 1×10^6 spermatozoa for 18h at 38.5 °C in a humidified 5% CO_2 incubator.

Applications of IVF

1. Research

The intended outcome of IVF is to produce a viable embryo, except when used as a means of detecting sperm capacitation or predictor of *in vivo* fertility. This technology, however, offers unprecedented opportunities for research to improve mammalian reproductive efficiency and to understand basic mechanisms involved in early development. IVF-derived embryos provide the raw material for comparative studies of gene expression for assessment of normal development. The transcriptomics and proteomics of IVF-derived embryos could be compared with their *in vivo* counterparts to understand genetic and enzymatic pattern of embryogenesis. IVF-derived embryos provide ample source of raw material to study and standardize such practices like microscopy, developmental rate, temperature sensitivity, freezability, embryo transfer, and effect of drugs, hormones and carcinogens, etc. on embryonic development. IVF-derived blastocysts are the closest to *in vivo* developed blastocysts and thus provide an *in vivo* embryonic model for research. Additional avenues of research are opening with further development of microinjection approaches building on current IVF technology, like bovine or buffalo ICSI. Much can be learned of fertilization, appropriateness of gamete preparation and of cytological events through further development of these potentially useful research tools. Bovine or buffalo IVF are used to model human IVF as they provide a non-ethical alternative for human IVF research. IVF-derived blastocysts are used for development of embryonic stem cell lines as well as trophoblast stem cells. IVF-derived embryos are used for studying the first differentiation event of inner cell mass and trophectoderm differentiation and later multi-lineage differentiation events like trophoblast formation, implantation and other embryonic events from zygote to blastocyst formation. IVF zygotes serve as starting tools to study and understand totipotency so as to have a sound understanding of processed like dedifferentiation and transdifferentiation.

2. Animal breeding

Bovine embryos resulting from IVF have already received much commercial interest and the same is expected of buffalo and small ruminant embryos in near future. Transvaginal oocyte retrieval provided a boon to commercialization of elite embryos. *In vitro* production of bovine embryos may enable replacement of currently practiced artificial insemination, since direct embryo transfer after cryopreservation should prove to be a more efficient means for pregnancy initiation. *In vitro* embryo production has been successfully combined with embryo sexing and embryo transfer in a commercial setting. It has been demonstrated that IVF procedures can effectively replace conventional *in vivo* embryo production methods when a predetermined number of pregnancies of known sex are needed within a short interval of time. IVF also provides a means of decreasing generation interval, to overcome infertility, to expand reproductive potential of an animal, to extend reproductive life, to assist in propagation of endangered cattle breeds, to produce large number of half-siblings simultaneously, to extend valuable semen *via* sperm injection, to assess gamete performance, to provide pronuclear ova for DNA microinjection, to provide the framework for a variety of gamete manipulations including cytoplasmic transfer, nuclear transfer and cloning by blastomeric recycling. IVF also extends reproductive life of a bull to decades by using the cryopreserved semen for IVF-embryo production. IVF has been used for prediction of fertility of animals. For example, the technique has been successfully employed to predict porcine fertility by homologous *in vitro* penetration of immature oocytes which enables discrimination between boars used for AI, characterized as low (<20%), intermediate (40–60%) and high (>80%) fertility groups.

3. Preservation of endangered species

In the past 200 years, more than 50 mammalian species have vanished and over 200 species are currently being threatened by extinction. The need for extending IVF to scarce gametes of endangered species is obvious but species differences have imposed major technological barriers. The potential for IVF for conservation of endangered mammalian species is nonetheless tremendous. Success has been attained using IVF to produce offspring in the Indian desert cat and in the Siberian tiger as well as several species of non-human primates. Progress in understanding fertilization mechanisms can be obtained through heterologous IVF, as was recently demonstrated with spermatozoa from endangered African antelope, the scimitar-horned oryx, and domestic cow oocytes.

Additional parameters and optimization for oocyte maturation, sperm capacitation, IVF and embryo culture for several species for complementing technologies that would provide great impetus for acceleration of additional refinements in IVF to afford better ways to enhance reproductive efficiency and to understand physiological events at the molecular level. The facility to produce embryos by combining previously cryopreserved gametes will be useful in breeding of laboratory, domestic and zoo animals. Refinements in animal IVF systems promise better ways to test for contraceptive development and models for improving assisted human reproduction.

Further reading

Austin C. (1951). Observation on the penetration of sperm into the mammalian egg. Australian Journal of Science Research, B4: 581–589.

Bigler D, Chen M, Waters S and White J. (1997). A model for sperm-egg binding and fusion on ADAMs and integrins. Trends in Cell Biology, 7: 220–225.

Brackett B, Hall J and Oh Y. (1978). *In vitro* fertilizing ability of testicular, epididymal and ejaculated rabbit spermatozoa. Fertility and Sterility, 29: 571–582.

Brackett BG, Keefer CL, Troop CG, Donawick WJ and Bennett KA. (1984). Bovine twins resulting from *in vitro* fertilization. Theriogenology, 21:224.

Brackett BG, Oh YK, Evans JF and Donawick WJ. (1980). Fertilization and early development of cow ova. Biology of Reproduction, 23:189–205.

Chauhan MS, Sigla SK, Palta P, Manik RS and Madan ML. (1998). *In vitro* maturation and fertilization, and subsequent development of buffalo (*Bubalis bubalis*) embryos: effect of oocyte quality and type of serum. Reproduction Fertility and Development, 110: 173–177.

Gasparrini B. (2007). *In vitro* embryo production in buffalo: Current situation and future applications. Italian Journal of animal Sciences, 6: 92–101.

Iritani A and Niwa K. (1977). Capacitation of bull spermatozoa and fertilization *in vitro* of cattle follicular oocytes matured in culture. Journal of Reproduction and Fertility, 50:119–121.

Moore K and Thatcher W. (2006). Major advances associated with reproduction in dairy cattle. Journal of Dairy Sciences. 89: 1254–1266.

Parrish JJ, Krogenaes A and Susko-Parrish JL. (1995). Effect of bovine sperm separation by either swim-up or Percoll method on success of *in vitro* fertilization and early embryonic development. Theriogenology, 44:859–869.

Pawshe C and Totey S. (1993). Effects of cumulus cells monolayer on *in vitro* maturation of denuded oocytes of buffalo (Bubalus Bubalis). Theriogenology, 39: 281.

Singh B, Chuahan MS, Singla SK, Gautam SK, Verma V, Manik RS, Singh AK, Sodhi M and Mukesh M. (2009). Reproductive biotechniques in buffalo (*Bubalus bubalis*): Status, prospects and challenges. Reproduction, Fertility and Development, 21: 499–510.

Tajik P and Niwa K. (1998). Effects of caffeine and/or heparin in a chemically defined medium with or without glucose on *in vitro* penetration of bovine oocytes and their subsequent development. Theriogenology, 49:771–777.

Wassarman P, Jovine L, Eveline S and Litscher. (2001). A profile of fertilization in mammals. Natural Cell Biology, 3: E59–64.

Chapter 9

In Vitro Embryo Culture

Introduction

Mammalian development begins with the fusion of haploid sperm and egg resulting to formation of 1-celled zygote. *In vitro* embryo production (IVEP) following IVM and IVF has gained tremendous importance for production of buffalo embryos as an alternative to multiple ovulation and embryo transfer (MOET) due to its advantages and flexibility, especially for the poor response of buffalo to hormonal treatments required for superovulation. However, the lack of knowledge concerning buffalo embryo physiology and reduced viability of cultured embryos limits the usefulness of this technique for genetic improvement of buffalo populations which currently ranks high in many Asian and Mediterranean countries where buffalo is reared as an economically important livestock species. In this respect, the ability to produce large number of embryos from donors of high genetic merit has considerable potential in disseminating genetic improvement and reducing the generation interval for rapid multiplication of elite germ plasm. The interest in buffalo breeding has observed a tremendous rise worldwide in the last few years due to critical role the species plays in many climatically disadvantaged agricultural systems.

Preimplantation development

Preimplantation development comprises the initial stages of mammalian development, before the embryo implants into the uterus. Beginning with fertilization, it proceeds as the embryo travels along the oviduct towards the uterus. During this time, the zygote divides sequentially to form a 2-, 4-, 8-, 16-cell embryo and so on. The division of the cells in specifically known as cleavage and each cell so formed is called as blastomere. It lasts for 4–7 days in most mammalian species. The cleavage rate in mammalian cells is among the slowest in animal kingdom, taking about 12–24 h. In contrast to radial cleavage of echinoderms and amphibians, mammalian cleavage is rotational. In mammals, the first cleavage is a normal meridonal division, while in second cleavage one of the blastomers divides meridonally and the other divides equatorially. Mammalain cleavage is also asynchronous in which all the blastomers do not divide at the same time. Thus, mammalian embryos do not increase exponentially from 2- to 4- to 8-cell stages, but frequently contain odd number of cells. At 8–16 cell stage, embryo shape resembles to blackberry and is therefore called as morula (Latin for mulberry). The morula consists of small group of internal cells surrounded by a larger group of external cells. The preimplantation embryo is covered by a protein shell called Zona pellucida that prevents embryo from attaching to the oviductal wall and causing an ectopic pregnancy. This is followed by "compaction" during which process the cells of the embryo become tightly attached and the morula adopts a more spherical and smoother shape. The blastomers huddle together, maximize their contact with one another and form a compact ball of cells. This tightly packed arrangement is stabilized by tight junctions that form between the outside cells of the ball, sealing of the inside of the sphere. The cells within the sphere form gap junctions, thereby enabling small molecules and ions to pass between them. This is followed by cavitation when blastomers accumulate fluid in the intercellular spaces to form a cavity, forming the blastocyst. The plasma membrane of the trophoblast cells contains a sodium pump (a Na$^+$ / K$^+$ - ATPase) facing the blastocoel. These pump sodium ions into the central cavity and the accumulation of these ions draws water osmotically, thus enlarging the blastocoel. The blastocyst grows as the cells divide and the cavity expands until it arrives to the uterus where it "haches" out from the Zona pellucida (hatching) and gets implanted to the uterine wall. It has been found that hatching occurs due to action of a trypsin-like protease, strypsin, which is located on the trophoblast cell membrane. It lyses a hole in the fibrillar matrix of Zona through which the expanding blastocyst escapes out. Once out, the blastocyst is caught by the uterine endometrium on its extracellular matrix of collagen, laminin, fibronectin, hyaluronic acid and heparin sulphate receptors. The trophoblast cells contain integrins that bind to the uterine collagen, fibronectin and laminin. They also synthesize heparan sulphate proteoglycan prior to implantation. Once in contact with the endometrium, the trophoblast secretes another set of proteases like collagenase, stromelysin and plasminogen activator. They digest the extracellular matrix of the uterine tissue, enabling the blastocyst to bury itself within the uterine wall. This is followed by development of a series of fetal and extraembryonic membranes (chorion, allantois and yolk sac) that mediate the exchange of gases, nutrients and waste products to and from the embryo. In mammals the chorion and the allantois form the placenta to allow the exchange to happen between the mother and embryo. Chorion also secretes hormones that cause the mother's uterus to retain the fetus and produce regulators of the immune response so that the mother will not reject the embryo. Thus, during preimplantation period, three cell populations are produced that make up the blastocysts: *epiblast* - the cells that make the fetus; *trophectoderm* and *primitive endoderm* or *hypoblast* – the two epithelia that form most of the extraembryonic membranes. The trophectoderm mediates implantation of embryo into uterus and formation of placenta, while primitive endoderm produces tissues that determine antero-posterior axis of fetus and form yolk sac.

As morula grows into a blastocyst, some of the cells remain on outside of embryo, while others are located on inside of morula forming an inner cell mass (ICM). The cells on surface develop asymmetrical regions on their plasma membrane, the first step towards epithelialization. The Apical surface is exposed to the outside medium, whereas lateral and basal (basolateral) surfaces are in contact with neighboring cells. The border between apical and basolateral surfaces is intercellular junction comprising of tight and adherens junctions, that hold cells together. The apical membrane is covered with microvilli and has associated proteins such as ezrin (associated with cytoskeletal protein actin) or atypical protein kinase C (aPKC). The basolateral membranes have E-cadherin, the main adhesion protein at this stage. The different expression of molecules on either side of cell surfaces, therefore, makes them polarized. The apical cells express regulatory transcription factors such as TEAD4, CDX2 or GATA3 which in turn direct expression of genes necessary for functions of trophectoderm. The expression of these regulatory transcription factors is thought to be induced by some of the proteins on apical surface of these cells, implying that position of cells in morula is a trigger for differentiation of cells into trophectoderm or ICM.

The cells of ICM do not have asymmetrical cell surfaces and thus are not polarized. These cells express NANOG and OCT4 genes and remain totipotent. They can differentiate into any of the cells in the embryo and even form trophectoderm if their position in the embryo is changed experimentally.

The cells of ICM differentiate into epiblast and extraembryonic primitive endoderm. Epiblast is considered as a pluripotent population and forms all the tissues of the fetus proper as well as other extraembryonic membranes, while primitive endoderm cells form an epithelium between epiblast and blastocyst cavity. It has been proposed that some ICM cells secrete fibroblast growth factor 4 (FGF4) very early on. This signals differentiation into epiblast and primitive endoderm cells. ICM cells that perceive this signal suffer a change in their gene expression and differentiate into primitive endoderm. The cells that produce FGF4 bear very few molecules of FGF receptors on their surface which makes them insensitive to it and therefore do not differentiate, thus forming the epiblast. The epiblast and primitive endoderm cells make up ICM of early and mid blastocyst. These two cell populations are intermixed in ICM in a "*salt and pepper*" pattern. As blastocyst matures, cells become separated into two populations in late blastocyst. The primitive endoderm cells form an epithelium between the epiblast and blastocyst cavity, while epiblast is encapsulated between trophectoderm and primitive endoderm.

Formation of extraembryonic membranes

Extraembryonic cells of the mammalian embryo form distinct mammalian tissues that enable fetus to survive within maternal uterus. The initial trophoblast cells divide normally and give rise to a population of cells wherein nuclear division occurs in absence of cytokinesis. This original type of trophoblast cells constitute a layer called cytotrophoblast, where as multinucleated type of cells form syncytiotrophoblast. Cytotrophoblast initially adheres to endometrium through a series of adhesion molecules. They also contain proteolytic enzymes that enable them to enter the uterine wall and remodel uterine blood vessels so that maternal blood bathes fetal blood vessels. Syncytiotrophoblast tissue is thought to help further progression of embryo into uterine wall by digesting uterine tissue. The uterine blood vessels eventually contact syncytiotrophoblast, followed by joining of extraembryonic mesoderm (from yolk sac/ hypoblast) to trophoblastic extensions forming blood vessels that carry nutrients from mother to embryo. The narrow connecting stalk of extraembryonic mesoderm that links embryo to trophoblast eventually forms vessels of umbilical cord. The fully developed extraembryonic organ, consisting of trophoblast tissue and blood vessel-containing mesoderm is the chorion. It fuses with uterine wall to form placenta. Chorion may be very closely apposed to maternal tissues, while still being readily separable from them (contact placenta of pig) or it may be intimately integrated with maternal tissues and the two cannot be separated without damage to both, mother and developing fetus (deciduous placenta of most mammals).

Parental contribution to embryonic development

It has been verified beyond any doubt that both the maternal and paternal genomes are required for successful embryonic development. At gamete fusion, fertilizing sperm DNA is packaged with protamines and mature sperm are transcriptionally inert. The sperm provides DNA for male pronucleus and is essential for egg activation. The mitochondria, microtubule organizing center precursors and the stored cellular components in sperm play minor roles in fertilization and early embryogenesis. Thus, it becomes essential for egg to provide suitable environment for sperm-egg recognition, prevention of polyspermy, parental genome remodeling and embryonic genome activation to ensure a successful transition from maternal control to a shared responsibility with male genome in directing early development. Maternal effect genes have been shown to affect multiple processes like pronuclear formation and fusion, the first cell division, embryonic gene transcription and cleavage-stage embryogenesis. Since oocyte is transcriptionally and translationally inactive during meiotic maturation and embryonic genome is not robustly activated until 2-cell stage, thereby implicating the role of maternal stored proteins in carrying out later stages of oocyte maturation, fertilization and first cleavage divisions. Thus, embryonic development is under the control of maternal genome for up to few divisions, after which switch from maternal to zygotic control occurs. This is referred to as *Maternal-to-Zygotic transition* (MZT) and occurs as early as at 2-cell stage in mouse and goat, and at 8-cell stage in most of the farm animals. The maternal to zygotic transition depends heavily on stored maternal components to form structures necessary to initiate development, eliminate redundant maternal materials and activate embryonic genome to reprogram gene expression during the MZT.

In vitro embryo culture

Mammalian embryos have been cultured in a variety of defined and undefined media. Four important events occur during *in vitro* culture period of bovine or buffalo embryos: first cleavage, activation of embryonic genome at 8- to 16- cell stage, compaction of morula on day 5 and blastocyst formation accompanied by formation of first two embryonic cell lines (ICM and trophoblast) (Figure 9) . Any or all of these events could be affected by inadequate culture conditions and therefore exert negative influence on blastocyst quality. It has been suggested that blastocyst yield after *in vitro* embryo production is mainly affected by intrinsic quality of oocytes, whereas blastocyst quality is determined by culture environment after fertilization. Hence, quality of *in vitro* produced blastocysts differs not only among different culture systems but also from *in vivo* produced embryos.

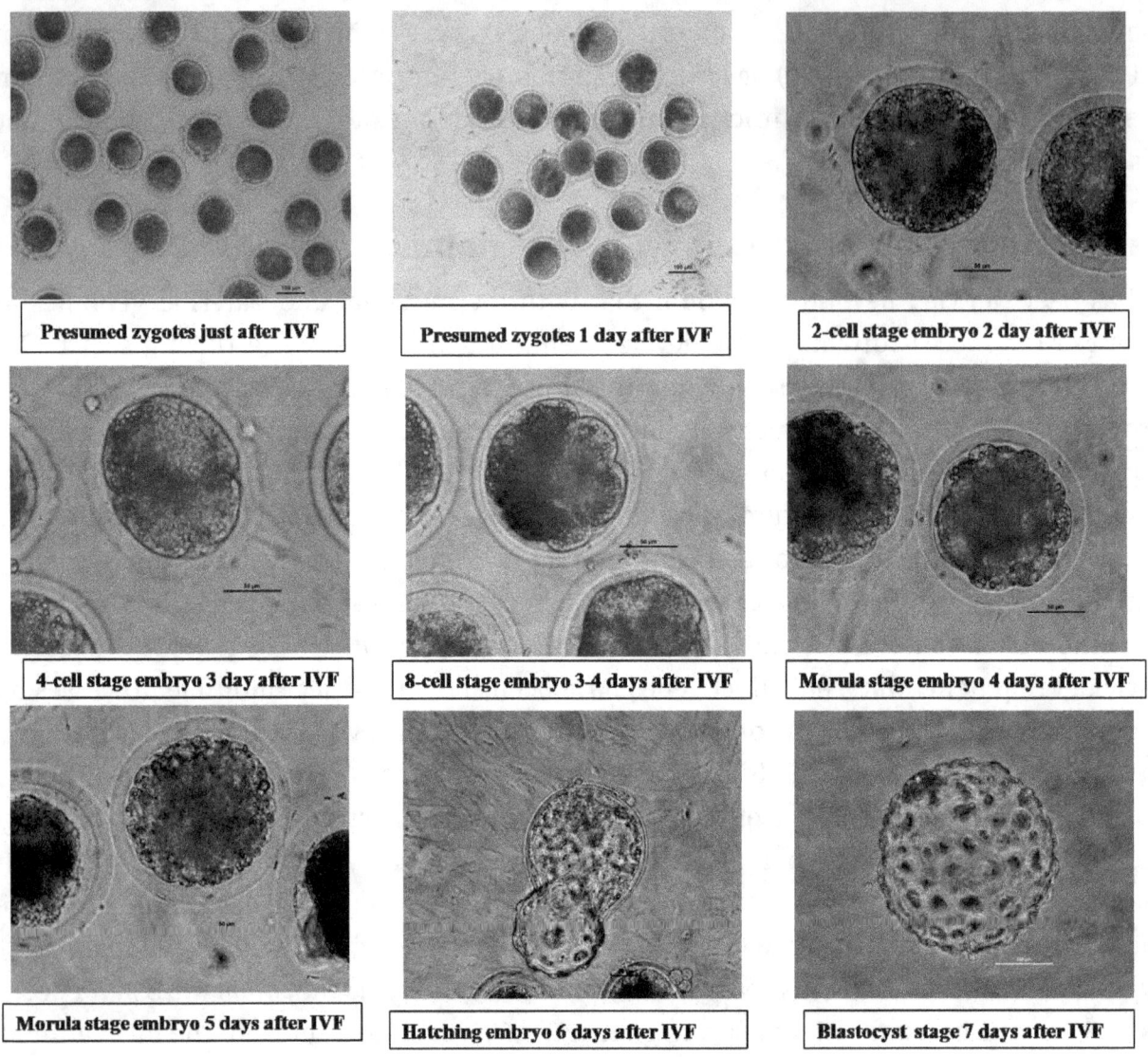

Fig. 9.1: In vitro embryo development in buffalo

It is being well understood that *in vitro* conditions cannot mimic dynamic changes of oviduct and uterus secretion that respond to the varying metabolism of a developing embryo. The optimal conditions for embryonic growth up to blastocyst stage are provided only by physiological environment of the preimplantation embryo. Lack of knowledge concerning buffalo embryo physiology and reduced viability of cultured embryos limits usefulness of assisted reproductive technologies for genetic improvement of buffalo populations. There have been substantial efforts to optimize culture conditions for *in vitro* production of buffalo embryos. The use of simple, defined culture media has made it possible to examine embryo's requirements for specific energy substrates such as glucose, pyruvate, lactate and glutamine.

Factors affecting *in vitro* embryo culture

The culture of embryos *in vitro* requires an appropriate environment so that an early embryo can undergo several cleavage divisions, enabling it to reach to blastocyst stage of development. *In vitro* embryo production is useful for: i) studying, in detail, the requirements of embryonic development; ii) mechanism and epigenetics of maternal-to-zygotic transition; iii) differential gene expression during development; and iv) cloning and transgenic animal production. The cleavage rate and embryonic development could be affected by many factors, right from oocyte quality, maturation, fertilization as well as post-fertilization culture conditions. The factors affecting oocyte quality and yield during *in vitro* maturation and *in vitro* fertilization have already been discussed. This section is devoted to factors affecting culture of fertilized oocytes up to blastocyst stage. At the end of the sperm-oocyte interaction, *in vitro* culture (IVC) is carried out, where presumptive zygotes are cultured *in vitro* in a culture medium at 38.5°C in a CO_2 incubator for up to 9–10 days to produce blastocysts. IVC is perhaps the most important step because of: i) Its longer duration of around 8–10 days compared to that of 24 h for IVM and 6–24 h for IVF; ii) the culture conditions and environment during IVC which have a profound influence on outcome of embryo development. Under *in vitro* conditions, buffalo embryos develop approximately 12–24 h earlier than cattle embryos. On day 6 of IVC it is possible to find embryos in advanced stages of development including hatched blastocysts but most embryos reach the blastocyst stage on day 7. A small proportion of embryos are delayed, reaching the blastocyst stage on day 8 but their quality and viability is poor.

IVC has been performed in a variety of chemically defined and undefined media. A chemically defined medium is a liquid prepared from bench chemicals and containing four or less basic components, *viz.*, inorganic salts, amino acids, vitamins and an energy source at known concentrations, while undefined medium is a liquid containing any biological fluid where the composition and the components vary considerably. The early embryo studies utilized undefined media such as blood plasma, blood sera, follicular fluid and chick egg extracts for culture.

1. Use of chemically defined culture media

IVC has been carried out effectively in chemically defined media supplemented with serum. Numerous laboratories using culture media without somatic cell support have reported embryo development results equal to, and in some cases better than those reported with co-culture. Serum has been found to stimulate morula compaction and blastocyst formation. Rosenkrans and First (1991) concluded that a simple medium, Charles Rosenkrans 1 (CR1), that contains essential and non-essential amino acids was beneficial to bovine embryo development *in vitro* in absence of feeder cells. It has been further shown that one-cell bovine or buffalo embryos can be successfully cultured beyond 8–16 cell block to blastocyst stage, using a chemically defined medium without glucose but containing pyruvate, lactate, amino acids and bovine serum albumin (BSA). The presence of glucose in IVC medium has been reported to inhibit early development, especially to 8-cell stage. However, addition of glucose at day 5 after IVF improves development to the blastocyst stage. In contrast to cattle and sheep, it has been demonstrated that presence of glucose is absolutely required for *in vitro* culture of buffalo embryos, particularly during the early embryonic development.

The most common protein source in chemically defined IVC media is serum or BSA. It has been demonstrated that serum-supplemented IVC medium has a biphasic effect, inhibitory at the first cleavage stage and stimulatory at the morula/blastocyst stage, on bovine embryo development *in vitro*. The serum was found to be most beneficial 2 days after insemination. However, serum substitution with PVP has shown lower blastocyst yield.

2. Co-culture with oviductal cells

Oviduct provides the microenvironment for transport and final maturation of gametes, fertilization and early embryonic development. Almost all species have been found to exhibit a block in embryonic development *in vitro*. For example, mouse and hamster exhibit block at 2-cell stage, porcine at 4-cell stage, and sheep, cow and buffalo at 8- to 16- cell stage of development. It has been found that embryos of these species stay within oviduct during that critical period when block to development occurs *in vitro*. This indicates that: i) oviduct and not the culture medium contains factors or conditions conducive to early embryonic development, and ii) certain developmental events occurring between 1 and 16-cell stage require specific environmental factors or conditions mainly provided by oviduct. The stage of blocked development occurs at a time of prolonged cell cycle, DNA synthesis and transition from maternal to zygotic control of development for mouse, cow, buffalo and pig. It is difficult to rescue embryo once the block has initiated. One of the widely used approaches to circumvent this block is to define optimum conditions for culture of farm animal embryos in terms of media and additives, gas atmosphere and stages of development of embryo. Currently, oviductal epithelial cells are commonly used for co-culture of cow, sheep, goat, pig and equine embryos. We found that simple medium like CR1aa supports buffalo embryonic development very well without oviductal cell co-culture or with co-culture on cumulus-granulosa cell bed of IVM droplets.

3. Culture of feeder layers

One of the approaches to overcome developmental arrest *in vitro* is to culture the embryos in presence of other cell types (feeder cells). These cells are thought to provide stimulus for development of the embryo. The feeder cells used for embryo co-culture fall into 3 groups: i) feeder cells from reproductive tissues; ii) feeder cells from non-reproductive tissues; iii) trophoblastic vesicles. The use of feeder monolayers for *in vitro* culture of farm animal embryos has not been reported prior to 1980's. Kuzan and Wright (1981) reported the first study on porcine blastocyst development, testing fibroblast cells from bovine uterine or testicular tissues as feeder monolayers. They found that uterine fibroblast monolayers provided a superior substratum for blastocyst attachment and trophoblastic outgrowth of porcine embryos *in vitro*. They also found that bovine morula hatched equally well on bovine uterine and testicular fibroblast monolayers. The use of bovine fetal uterine fibroblast cells for culture of bovine and equine embryos has proved superior over that of medium alone. The positive effect exerted by reproductive tract cells may be non-specific as other types of somatic cells (skin, testis, liver) from different species also support the development of bovine embryos *in vitro*. These co-culture systems have been criticized for their undefined nature but they do achieve the desired goal of circumventing the *in vitro* block and producing viable embryos. The feeder cells are thought to provide some factor (s) to the embryos that improve their *in vitro* development. This "*positive conditioning*" may involve a growth factor or some other factor normally produced by such cells or they might be removing an inhibitory component from the culture environment, i.e., "*negative conditioning*" or both positive and negative conditioning may be occurring simultaneously. The use of these co-culture systems have proved beneficial in circumventing the development block in IVM/IVF bovine and buffalo embryos but the proportion of these embryos reaching morula/ blastocyst stage was found to be small (25–40% of inseminated oocytes) and only a few hatched. This may be due to some fundamental underlying problem, like inadequacy of the culture conditions or possibly abnormal oocyte maturation which demands further studies to delineate the actual cause.

4. Use of pre-conditioned media

A number of researchers have used media pre-conditioned by exposure to oviductal/ uterine or cumulus/granulosa cells for culture of farm animal embryos. The medium conditioned by oviductal tissue was found to be effective as co-culture in supporting bovine embryo-development from the zygote to blastocyst stage. The supernatants prepared from cells taken around estrus were significantly more effective than those prepared at other times. It is apparent that the conditioned medium contains low molecular weight factor (s) acting in early cleavage and larger molecule (s) acting in blastocyst formation. The advantages of serum-free conditioned medium over co-culture include eliminating the confounding presence of cells and making the search for soluble embryotrophic factors easier. It was reported that addition of high molecular weight bovine oviduct conditioned medium (BOCM) fraction to modified synthetic oviductal fluid (mSOF) medium significantly increased embryo development up to blastocyst stage compared to mSOF and BOCM.

5. Hormones and growth factors

Insulin has been shown to increase the rate of glucose transport in blastocyst and blastocyst metabolism *in vitro*. Insulin is considered as the only hormone shown to have a clear effect on early embryonic development. The beneficial effect of insulin on IVC are thought to be mediated *via* IGF-I receptors.

Growth hormone (GH) has also been reported to show beneficial effects on cleavage rate, blastocyst formation and hatchability of the embryos. It has been suggested that a functional GH receptor capable of modulating carbohydrate and lipid metabolism synthesis during early preimplantation development of bovine embryos is subject to activation by embryonic GH.

6. Growth factors

IGF-I has been shown to stimulate protein synthesis, mitogenesis of inner cell mass and morphological development *via* insulin and IGF-I receptors. However, some studies showed no evidence of improvement in bovine embryo development up on IGF-I supplementation. Epidermal growth factor (EGF) has also been shown to be capable of significantly improving the development of 2- to 8- cell cattle embryos to the morula/ blastocyst stage. TGF-β supplemented in CR1aa medium was also found to have a positive effect on improving bovine and buffalo embryonic development. Human luekaemia inhibitory factor (hLIF) supplementation also improves the development of cattle and buffalo embryos when added to SOF medium supplemented with BSA or PVP but not with human serum.

7. Frequency of medium change

In order to reduce the accumulation of free radicals, ammonium and other catabolites that may affect embryo development, it has been suggested to use the easy expedient to change the medium more times during culture. However, no significant differences in buffalo embryo development have been recorded by changing the IVC medium 3 (day 1, 3 and 5) or 2 (day 1 and 5) times during culture, with a tendency of improvement in latter case. Therefore, in contrast to other species, the addition of fresh medium on day 3 of culture in buffalo does not exert any positive influence. It is likely due to higher sensitivity of buffalo embryos to fluctuations of temperature and/or pH that normally occur during culture change. Thus, buffalo embryos should not be disturbed during the culture. We found comparable embryo development rate when presumed zygotes were cultured in CRaa1 and Research Vitro Cleave medium (K-RVCL-50; Cook, Brisbane, Qld, Australia) supplemented with 1% fatty acid-free BSA. The advantage of the latter was the embryo culture medium was least disturbed as it does not require media changes up to 7 to 9 days post insemination.

8. Amino acid supplementation

It has been reported that development of early cleavage stages is stimulated by nonessential amino acids and glutamine supplementation, while development beyond day 3 post-insemination is stimulated by a combination of nonessential amino acids (1%) and glutamine (1mM). The reduction of essential amino acid concentration is suggested to be beneficial to culture of buffalo embryos. Apart from their role in protein synthesis, amino acids have additional role as osmolytes, intracellular buffers, heavy metal chelators, energy sources as well as precursors for versatile physiological regulators such as nitric oxide and polyamines.

9. Miscellaneous factors

i) Buffering system and osmolarity

The buffering system employed for IVEP depends on whether the medium is exposed to air or to carbon dioxide- enriched atmosphere. IVEP is routinely performed in 5% CO_2 incubator. HEPES or phosphate-buffered media could be used for short-term work with oocytes or embryos if 5% CO_2 is not available. The osmolarity of the media for IVEP should be between 275–285 mOsm.

ii) Water quality

Water is major constituent of IVEP media. The use of ultrapure water, free from contaminants is crucial. The principal methods for obtaining purified water are glass distillation, deionization, filtration, reverse osmosis, adsorption and ultra-filtration. Optimum results have been reported using Millipore reverse osmosis (RO) and Milli-Q (MQ) water.

iii) Temperature and gas phase

The optimum temperature for buffalo IVEP is 38–39˚C. 40˚C is detrimental to fertilization and early development. It is worth mentioning that exposure of oocytes to 20˚C during recovery decreases the percentage of oocytes that undergo fertilization and subsequent development *in vitro* and induces chromosomal abnormalities. It is reported that 5% O_2, 5% CO_2 and 90% N_2 gas mixture provides a suitable atmosphere for early bovine and buffalo embryo growth *in vitro*. However, most of the researchers, including the authors, recommend the gas phase of 5% CO_2 in air and 38.5˚C incubation temperature for buffalo IVEP.

iv) Effect of light

The general principle should always be to culture embryos in darkness and not in light.

v) Protection from oxygen toxicity

The exposure to 20% oxygen and light during routine embryo manipulation may lead to generation of superoxide radicals. The elimination of these radicals from embryo culture has been found to result in significant improvement in embryo development.

Further reading

Wright R, Anderson G, Cupps P and Drost M. (1976). Successful culture *in vitro* of bovine embryos to the blastocyst stage. Biology of Reproduction 14: 157–162.

Gasparrini G. (2007). *In vitro* embryo production in buffalo: current situation and future perspectives. Italian Journal of Animal Sciences, 6: 92–101

Badr M. (2009). Effects of supplementation of amino acids on *in vitro* embryo development in defined culture media. Global Veterinaria, 3: 407–413.

Nandi S, Raghu HM, Ravindranatha BM, Chauhan MS. (2002). Production of buffalo (Bubalus bubalis) embryos. Reproduction in Domestic Animals, 37: 65–74.

Abdoon A, Kandil O, Otoi T, Suzuki T. (2001). Influence of oocyte quality, culture media and gonadotropins on cleavage rate and development of *in vitro* fertilized buffalo embryos. Animal Reproduction Science, 65: 215–223.

Apa R, Larnzone A, Miceli F, Masrrandrc M, Caruso A, Mancuso,S. and Canpiari R. (1994). Growth hormone induces *in vitro* maturation of follicle and cumulus-enclosed rat oocytes. Molecular Reproduction and Development, 106: 207–212.

Augastin R, Pocar P, Navarretae-Santos A, Wrenzycki C, Gandolfi F, Neimann H and Fischer B. (2001). Glucose transport expression is developmentally regulated in *in vitro* derived bovine preimplantation embryos. Molecular Reproduction and Development, 60: 370–376.

Eyestone W, Jones J and First N. (1990). The use of oviduct-conditioned medium for culture of bovine oocytes to the blastocyst stage. Theriogenology 33: 226.

Gardener H and Kaye P. (1991). Insulin increases cell numbers and morphologyical development in mouse pre-implantation embryos *in vitro*. Reproduction Fertility and Development, 3: 79–91.

Rosenkrans C and First N. (1994). Effect of free amino acids and vitamins on cleavage and developmental rate of bovine zygotes *in vitro*. Journal of Animal Science, 72: 434–437.

Boland MP. (1984). Use of the rabbit oviduct as a screening tool for the viability of mammalian eggs. Theriogenology, 21: 126–137.

Appendix 4

In vitro culture of buffalo embryos

At the end of sperm-oocyte incubation, the cumulus cells are washed off from the oocytes by gentle pipetting/ vortexing. The oocytes are pipetted out of the debris under the microscope, then washed several times with modified Charles Rosenkrans medium with amino acids (mCR2aa) containing 0.6% Fraction V BSA and cultured in this medium in humidified CO_2 incubator (5% CO_2 in air) at 38.5°C on the original granulosa cell beds of *in vitro* maturation droplets. After 48h, half of the medium is replaced with IVC medium (mCR2aa + 0.6% BSA fraction V + 10% FBS) followed by two to three such media changes up to 7 or 8 days post fertilization, until hatched blastocysts develop.

In vitro culture media

Washing medium for presumed zygotes

mCR2aa	20 ml
BSA (Fraction V) @ 0.8%	0.16 gm
Gentamicin	50µg/ml

In vitro culture medium

mCR2aa	9.0 ml
FBS	1.0 ml
Gentamicin	50 µg/ml

During the preparation of above media, all the components are mixed well and incubated in a CO_2 incubator at 38.5°C, 5% CO_2 for 2 h for stabilization of pH and temperature before further use. The media are filtered through 0.22 µm filter just before.

Modified Charles Rosenkrans-2 medium with amino acids (mCR2aa)

The mCR2aa medium is prepared in 100 ml aliquots as per the compositions given below. The pH and the osmolarity of the media are checked each time a fresh lot of the medium is prepared.

Composition of mCR2aa

Component	Molarity (mM)	Component	Molarity (mM)
Water	-	Na Pyruvate	0.5
NaCl	108.3	Glycine	0.5
NaHCO$_3$	24.9	Alanine	0.5
NEAA*	1.0 ml/100 ml	Glucose	1
EAA**	2.0 ml/100 ml	Phenol red	5 µg/ml
Glutamine	1	Gentamycin	50µg/ml
KCl	2.9	BSA	0.6%
Hemicalcium lactate	2.5		

*MEM Non-essential amino acids (Sigma, Catalog No. M7145)

*BME Essential amino acids (Sigma, catalog No. B6766)

Chapter 10
Parthenogenesis

Introduction

Sexual reproduction is a complex process that involves two basic elements: i) Meiotic reduction - chromosomal segregation, assortment and crossing over that generate an immense variety of haploid gametes; and ii) Syngamy - fusion of gametes that produces unique new individuals in each generation. The mixing of genotypes from different individuals, recombination, is the essential characteristic of sex in eukaryotic organisms, and circumvention of these processes leads to parthenogenesis and cloning.

Modes of asexual reproduction

Parthenogenesis is distinct, but comparable, to vegetative reproduction (budding, fragmentation, fission etc). It is common in plants and some invertebrate animals in producing clones, as they do not involve egg production and meiotic processing of chromosomes. However, in ecological sense this vegetative mode of reproduction is comparable with growth than reproduction, given the fact that fertilized seeds and eggs are often the essential dispersal phase of many plants and animals, while in most cases the vegetative propagules tend to remain close to the parent organism.

Cyclical parthenogenesis is another form of reproduction which alternates between sexual and asexual egg production. The cyclical parthenogens engage in periodic recombination and thus are referred to as facultatively sexual. For example, *Daphina pulex* produces a new assemblage of clones after each cycle of sexual reproduction.

True parthenogenesis is a strictly clonal form of reproduction that transmits the female's diploid (or polyploid) genome to eggs, which develop spontaneously into genetically identical daughters. It can further be classified into *apomixs* or *automoxis*. Apomixis refers to zygote production without chromosome reduction. The reductional division (meiosis I) is eliminated and non-recombinant eggs with a single equational division (meiosis II) are produced. It leads to strict clonal inheritance and retention of maternal level of heterozygosity. Automixis is referred to as meiotic parthenogenesis and restores diploidy by fusion of meiotic products. In most cases, automixis is comparable to self-fertilization and quickly leads to complete homozygosity. Some automicts produce normal haploid ova and then duplicate the generative nucleus in a subsequent mitotic division. Fusion of these mitotic products restores diploidy but leads to complete homozygosity in one step. The inheritance is effectively clonal once automicts are completely homozygous. *Rhabditis* nematodes, for example, fuse the second polar body with the egg nucleus, leading to self-fertilization. Most of the parthenogenetic animals are functionally apomictic. They retain elements of meiosis while circumventing chromosomal recombination and reduction.

Sperm-dependent mode of parthenogenetic reproduction is seen in *Dandelions* of North America. They are psedogamous apomicts where pollination is necessary to activate development of endosperm tissue in the seed but the generative nucleus develops apomictically. Pseudogamy is more commonly called gynogenesis in animals. Despite the need for sperm, pseudogamous inheritance is strictly maternal and clonal. Although pseudogamous forms are not parthenogenetic in the strict sense, genetic consequences are the same. *Alsophila pometaria* has pseudogamous lineages that use sperm from males of a coexisting sexual lineage, while *Poecilia formosa* use sperm from males of a closely related sexual species.

Hybridogenesis is an unusual form of matrilineal inheritance that perpetuates a hybrid genotype. It combines elements of parthenogenesis and sexual reproduction. The hybrid fish *Poeciliopsis monacha-lucida* is a hybrid between the sexual species *P. monacha* and *P. lucida*.

Parthenogenesis in farm animals

Parthenogenesis is the growth and development of embryos out of oocytes that have not been fertilized by sperm. It occurs naturally in many invertebrates as well as in some vertebrates and can be induced chemically in mammalian oocytes. The produced embryos have been used widely to understand early development events as well as a source of embryonic stem cells. These embryonic stem cells have the ethical advantage of not involving the destruction of viable embryos but the limitation of being homozygous for most genes.

Mechanism of parthenogenesis

Ovulated mammalian oocytes are arrested at the metaphase II (MII) stage of meiosis and essentially depend upon sperm for completion of meiosis. Sperm is responsible for releasing oocyte from meiotic arrest and for inducing other events that are collectively referred to as *oocyte activation*. As discussed in the chapter on IVM, this oocyte activation includes cortical granule exocytosis, reinitiation of meiosis, extrusion of the second polar body, formation of pronuclei and recruitment of mRNA. In all mammalian species studied so far, oocyte activation is triggered by repetitive rises in the intracellular concentration of free Ca^{2+} which is regarded as the sufficient and indispensable event for activation. The rises are generated by release of Ca^{2+} from the intracellular stores which is mediated by production of inositol 1,4, 5-triphosphate (IP_3), following activation of the phosphoinositide signaling pathway. Upon fusion with the oocyte the sperm introduces a protein factor responsible for inducing production of IP_3 and Ca^{2+} release. Recent research suggests that the sperm factor is phospholipase C-zeta (PLCZ1). This PLC variant is sperm-specific and induces sperm-like Ca^{2+} oscillations when injected into mouse oocytes. PLCZ1 is localized to the postacrosomal region in mouse sperm, while in cattle it is detected at equatorial region of sperm. PLCZ1 catalyzes the hydrolysis of phosphatidyl 4, 5-bisphosphate (PIP_2), producing IP_3 and diacylglycerol (DAG). The elevation in IP_3 concentration is responsible for inducing Ca^{2+} release from the endoplasmic reticulum, upon binding its cognate receptor IP_3R-1, which is mostly located in this organelle. The continuous production of IP_3 is thought to underlie the persistence of the oscillations during mammalian fertilization and eventually lead to IP_3R-1 degradation. This IP_3R-1 down regulation is a hallmark of fertilization and is thought to contribute to decreased responsiveness to IP_3 observed after fertilization.

The experimental induction of parthenogenesis in mammals began with the pioneering studies of Pincus and his colleagues in rabbit. They found that extrusion of polar bodies could be induced *in vitro* not only by contact with sperm suspension but also by heat treatment or exposure to butyric acid and hypertonic solutions. The general procedure of parthenogenetic embryo development is almost similar to the *in vitro* development of fertilized oocytes except the step of parthenogenetic activation with different activation agents which could either be electrical, chemical or other types. Thus, the general procedure would include collection of ovaries, *in vitro* maturation, activation of oocytes with different activation agents and finally *in vitro* culture. Since, Ca^{2+} transient is the key trigger of meiotic resumption during fertilization, a wide range of procedures for artificial oocyte activation have been established including mechanical, chemical and physical stimuli that elicit one or several Ca^{2+} transients in the oocyte. It has been demonstrated that mechanical disruption of frog oocytes with a fine needle is sufficient to generate Ca^{2+} influx and initiation of development. Microinjection of Ca^{2+} has been found effective for porcine oocytes. Chemical activation can be induced by exposure to Ca^{2+} ionophore, 7% ethanol, strontium chloride, phorbol ester and thimerosal. Ionophore A23187 promotes the release of intracellular Ca^{2+} stores, in addition to facilitating the influx of extracellular Ca^{2+} ions. Ionomycin is another potent Ca^{2+} ionophore currently used in nuclear transfer/ cloning. It mobilizes intracellular Ca^{2+} by depletion of Ca^{2+} stores. Exposure of matured oocytes to 7% ethanol for 5–7 min induces successful activation and pronuclear formation by promoting the formation of IP_3 and the influx of extracellular Ca^{2+}. These substances induce a single Ca^{2+} rise in oocyte, while the initial Ca^{2+} rise is normally followed by Ca^{2+} oscillations during fertilization in mammals. Strontium chloride induces multiple Ca^{2+} transients probably by displacing bound Ca^{2+} in the oocyte but also by inducing intracellular Ca^{2+} release. Phorbol ester which mimics endogenous diacylglycerol activates the calcium-and phosphorous-dependent protein kinase C and induces calcium oscillations and pronuclear formation in mouse oocytes. Thimerosal, a sulfhydryl–oxidizing agent that indices repetitive Ca^{2+} oscillations has been successfully used for the activation of bovine oocytes. Electrical stimulation is an alternative to chemical activation to induce Ca^{2+} influx through the formation of pores in the plasma membrane. The success of this procedure depends on size of pores formed, and also on the ionic contents of the medium and cell type. Periodically repeated electrical stimulation mimics the pattern of oscillations observed during fertilization. The electrical stimulation is suggested to stimulate the production of IP_3 that leads to intracellular Ca^{2+} release. Exposure of oocytes to room temperature has also been used as a physical stimulus for oocyte activation prior to nuclear transfer. The next step for resumption of meiosis, chromatin decondensation and transition to interphase is the drop of MPF (maturation promoting factor) and MAPK activity. MPF is described as the biological activity of cytoplasm capable of reinitiating meiosis in prophase arrested oocytes and is essential for meiotic arrest at MII. Since the concentration of Cyclin B oscillates during the cell cycle, the level of MPF molecules depends on the synthesis and degradation of Cyclin B. MPF activity rises during oocyte maturation and phosphorylation status of its constituents determines its kinase activity. The phosphorylation is regulated by specific kinases, like cyclin-activating kinase, and other kinases (Myt-1 and Wee) as well as cdc25. The phosphatase cdc25 activates MPF by dephosphorylation of tyr15 and thr14 sites of cdc2. Specific substances that inhibit the activation site of this phosphatase are currently used for oocyte activation. During MII arrest, high MPF activity is maintained through continuous equilibrium between Cyclin B degradation and synthesis, explaining the role of protein synthesis inhibitors in inducing oocyte activation. MPF inhibition under laboratory conditions is achieved through a variety of approaches. One of the possible ways is to incubate matured oocytes in broad-spectrum inhibitors of protein synthesis and/or protein phosphorylation. Inhibitors of protein synthesis like cycloheximide and puromycin induce oocyte activation after prolonged periods of incubation in mouse and human oocytes, while the rat and pig oocytes are not activated by this treatment. The combined use of a Ca^{2+} stimulating substance with an inhibitor of protein synthesis has been widely used for activation of mouse, sheep and cattle oocytes. For example, Ca^{2+} ionophore plus cycloheximide induce a high rate of pronuclear formation and development to blastocyst stage of mouse oocytes. Ethanol plus cycloheximide have also been used successfully in bovine oocyte activation. Cycloheximide not only depletes the oocyte from proteins maintaining MPF activity but also inhibits the translation of proteins responsible for initiation of DNA replication. A more specific inhibition is achieved by inhibition of protein kinases. It has been shown that a cAMP-dependent kinase inhibitor is localized in the nucleus of G2/M cells. Upon its inhibition, the cycle is arrested suggesting that protein kinase inhibition is a normal function for the transition from M-phase to G1. For example, the treatment of *Xenopus laevis* oocytes are exposed to 6-DMAP, a phosphatase inhibitor, and resumption of meiosis without Ca^{2+} release is observed. Similarly, pig oocytes treated with staurosporine and H7 (protein kinase inhibitors) resume meiosis without Ca^{2+} release, and undergo pronuclear formation and development to blastocyst. However, mouse, bovine and buffalo oocytes are not activated when incubated in 6-DMAP without previous Ca^{2+} release.

The combination of a Ca^{2+} ionophore with 6-DMAP induces high rates of activation, pronuclear formation and development to blastocyst in ovine, bovine and bubaline species. These protocols have been successfully used for production of cloned calves after somatic cell nuclear transfer. However, oocytes activated with ionomycin and 6-DMAP display some alterations in the DNA content, reflecting an abnormal pattern of karyokinesis during the first cell cycle. Numerous failures in the establishment of pregnancies, placental malformations and perinatal death have been reported after activation of nuclear transfer embryos with this protocol. 6-DMAP inhibits phosphorylation of ribosomal protein S6 and activation of the 70 kDa S6 kinase in somatic cells and it drastically affects cytoskeletal components leading to formation of micronuclei containing chromosomes. This suggests that a disturbance in G1 of a signal transduction pathway may contribute to abnormal mitosis. 6-DMAP induces premature chromatin condensation and premature mitosis in cells arrested in S-phase, suggesting a role for protein dephosphorylation in control of mitosis. Protein kinase inhibition is an efficient way to induce oocyte activation. However, it should be considered that these inhibitors are not specifically interfering with one kinase but with several involved in other cell functions, whose inhibition may be deleterious in subsequent cellular events after activation. Among diverse protein kinases, two MAPK isozymes, extracellular signal-regulated protein kinases 1 and 2, have been found to be active during meiosis. Phosphorylated MAPK plays role in spindle formation and polar body diffusion or extrusion. MAPK activation is essential for MII arrest in mouse and bovine oocytes. The decrease in MAPK activity is correlated with the formation of a nuclear envelope after parthenogenetic activation of mouse and cattle oocytes as well as initiation of DNA synthesis. This suggests that inhibition of MAPK activity, independently of MPF inactivation, leads to oocyte activation.

Since, a coordinated series of events is responsible for the signaling pathway initiated by sperm during fertilization, it is reasonable to think that specific kinases and phosphatases are involved differentially in the transition from MII arrest to interphase. The cell cycle is regulated by kinases that are activated by cyclin binding and phosphorylation and inhibited by phosphorylation, proteolysis and binding to specific CDKIs. The inhibition of specific kinases by these molecules during oocyte activation is of high interest in a way to mimic sperm-mediated events during fertilization. The development of synthetic CDKIs in the last years has been continuously growing, since several human diseases can be treated with these compounds. The first compound identified as CDKI has been 6-DMAP, a substance of low selectivity. Butyrolactone I, isolated from *Aspergillus* strain 25799, selectively inhibits cdk2 and cdc2, arresting the cell cycle at G2/M and G1/S. By using combinatorial chemistry, new compounds have been synthesized with high selectivity and efficiency. These purine analogues target the ATP-binding site of cyclin/cdk molecules. Olomoucine, roscovitine and bohemine are some of these compounds with high selectivity.

The best parthenogenetic activation results have been obtained by use of intracellular calcium transient inducing agents combined with protein synthesis inhibitors (cycloheximide) or protein phosphorylation inhibitors (6-DMAP). Although cycloheximide and 6-DMAP allow high rates of parthenogenetic activation and development of bovine embryos reconstructed by nuclear transfer to the blastocyst stage, high proportions of advanced pregnancy losses, perinatal death and severe abnormalities in the fetal and neonatal conceptus development, suggest that these activation protocols may have detrimental effects on embryo development. In one of the study aiming to evaluate embryonic development in different grade bovine oocytes upon treatment with different chemicals like A23187 + 6-DMAP; Ionomycin + 6-DMAP;

Ethanol + 6-DMAP; and Ionomycin + Cycloheximide; Ionomycin + 6-DMAP treatment, resulted to highest embryonic development rates for 8-cell, morula and blastocyst stages. However, no consistent relationship was observed between oocyte grades and embryo development rates.

Genomic imprinting of parthenogenetically activated embryos

When the mammalian oocyte is fertilized with sperm, it receives the paternal genetic material. The paternal alleles, like the oocyte alleles, are subjected to epigenetic modifications during gametogenesis that cause a subset of mammalian genes to be expressed from one of the two parental chromosomes in the embryo. This regulatory mechanism is termed as *genomic imprinting*. Additional epigenetic modifications also occur during early development after fertilization. In this context maternal and paternal genomes are not functionally equivalent implying the need of both the genomes in a developing embryo for normal mammalian development. It has been noted that mammalian parthenotes are able to undergo several cycles of cell division after oocyte activation but seldom proceed to term, arresting at different stages of development, depending on the species. The reason for the arrested development is believed to be abnormal genetic imprinting as all genetic material in parthenotes is of maternal origin, thereby preventing proper development of extraembryonic tissues whose expression is regulated by the male genome. Uniparental embryos, such as parthenotes and androgenotes, have been used to study imprinting processes as well as the role paternal genome plays during early embryonic development. Since, diploid parthenotes (DPs) and fertilized embryos show similar development, at least up to blastocyst stage, their gene transcription patterns during early developmental processes may not differ markedly. However, subtle differences may exist due to expression of Y-chromosome linked genes and imprinting genes during early development, unlike DPs. Compared to DPs, fewer haploid parthenotes (HPs) cultured *in vitro* reach to blastocyst stage. The reasons for this limited developmental potential are not clear, though this could be due to lack of genetic component (s) in HPs which might increase the duration of cell cycle and consequently slow their development. Also low DNA content in HPs may not be sufficient to control the gene expression network which could result in apoptosis or the failure of developmental processes during preimplantation development. It has been demonstrated that fertilized blastocysts express several genes at higher levels than DP blastocysts, like eukaryotic translation initiation factor 2 (Y-chromosome specific gene), subunit 3, structural gene Y-linked (Eif2s3Y) and the imprinting gene U2, small nuclear ribonucleoprotein auxillary factor 1 and related sequence 1(U2af1-rs1).

Genomic imprinting plays important roles in the regulation of fetal growth, development, placental function and post-natal behavior. It endows some genes with different "imprints" which leads to their differential expression in fetus and/or placenta and regulate transfer of nutrients to fetus and placenta from mother.

Genomic imprinting is controlled by DNA methylation, histone modifications, noncoding RNA and specialized chromatin structure. DNA methylation in differentially methylated regions (DMRs) of parental origin allows discrimination between maternal and paternal alleles and leads to monoallelic expression of imprinted genes. Uniparental fetuses, including parthenotes and androgenotes, show disrupted expression of several imprinted genes such as Snrpn, Peg3, H19 and Gtl2. It has been shown that paternal genome is more important for development of extrembryonic tissues, while the maternal genome is more essential for fetal development. Hence, aberrant epigenetic modifications caused by inefficient reprogramming severely undermines developmental potency of parthenogenetic embryos. It has been shown that maternally imprinted genes like Snrpn and Peg/Mest are nearly unmethylated or heavily methylated, respectively, in their DMRs at the 2-cell stage in parthenogenetic embryos, while at morula stage both the genes showed almost complete methylation of all CpG sites. Peg 3 also shows strong *de novo* methylation in parthenogenetic blastocysts. In contrast, the paternally imprinted genes H19 and Rasgrf1 show complete unmethylation of their DMRs at morula stage in parthenogenetic embryos. These findings indicate that diploid parthenogenetic embryos adopt a maternal-type methylation pattern on both sets of maternal chromosomes and that the aberrantly homogenous status of methylation imprints may account for developmental failure.

Despite of all the limitations and lack of knowledge, a significant leap was achieved in mammalian parthenogenetics with the successful birth of fatherless mouse, Kaguya, in 2004 in Japan. Kaguya is the first viable parthenogenetic mammal and might be extremely helpful in understanding the molecular processes involved during genomic imprinting which is the main barrier to parthenogenetic development in mammals.

Suggested reading

Graham CF. (1974) The production of parthenogenetic mammalian embryos and their use in biological research. Biological Reviews, 49: 399–422.

Paffoni A, Brevini T and Gandolfi F. (2008) Parthenogenetic activation: Biology and applications in the ART laboratory. Placenta, 29: S121-S125.

Pincus G and Shapiro H. (1940) Further studies on the parthenogenetic activation of rabbit eggs. Proceedings of the National Academy of Sciences of the United States of America, 26: 163–165

Chien-Tsung L. (2002) Parthenogenesis of rabbit oocytes activated by different stimuli. Animal Reproduction Science, 70: 67–276.

Parnpai R. and Tasripoo K. (2003) Effect of different activation protocols on the development of cloned swamp buffalo embryos derived from granulosa cells. Theriogenology, 59: 279.

Alberio R, Kubelka M, Zakhartchenko V, Hajduch M, Wolf E and Motlik J. (2000). Activation of bovine oocytes by specific inhibition of cyclindependent kinases. Molecular Reproduction and Development, 55: 422–432.

Brind S, Swann K Carroll J. (2000). Inositol 1, 4, 5-Trisphosphate receptors are downregulated in mouse oocytes in response to sperm or adenophostin A but not to increases in intracellular Ca(2+) or egg activation. Developmental Biology, 223: 251–65.

Kaufman MH. (1979). Mammalian parthenogenetic development. Reproduction 33: 261–264.

Allen ND, Barton SC, Hilton K, Norris ML, Surani MA. (1994). A functional analysis of imprinting in parthenogenetic embryonic stem cells. Development 120:1473–1482.

Balakier H and Tarkowski AK. (1976). Diploid parthenogenetic mouse embryos produced by heat-shock and cytochalasin B. Journal of Embryology and Experimental Morphology 35: 25–39.

Mehlmann LM. (2005). Stops and starts in mammalian oocytes: recent advances in understanding the regulation of meiotic arrest and oocyte maturation. Reproduction 130: 791–799.

Bell G. (1982). The Masterpiece of Nature: The Evolution and Genetics of Sexuality. Univ. of California Press, Berkeley.

Ross P, Beyhan Z, Iager A, Yoon S, Malcuit C, Schellander K, Fissore R and Cibelli J. (2008). Parthenogenetic activation of bovine oocytes using bovine and murine phospholipase C zeta. BMC Developmental Biology 8: 16.

Moore T and Ball M. (2004). Kaguya, the first parthenogenetic mammal-engineering triumph or lottery winner. Reproduction, 128: 1–3.

Appendix 5

Parthenogenetic embryo production

After 24 h of IVM, COCs with expanded cumulus are transferred into 1.5 ml microcentrifuge tube containing 500μL Hyaluronidase (0.5mg/ml) in T2 and incubated for 1 min at 38.5°C. It is followed by gentle pipetting using 1 ml microtip and subsequent vortexing at high speed for 2–3 min. The contents of the tube are transferred to a 35 mm dish containing mCR$_2$aa and completely denuded oocytes with evenly granular cytoplasm are selected and washed twice in fresh mCR$_2$aa for removal of residual cumulus cells. The denuded oocytes with a prominent polar body are parthenogenetically activated by exposure to 7% ethanol/ 5 μM calcium ionophore (A 23187) for 7 min and 5 min respectively, followed by incubation with 2 mM 6-dimethyl aminopurine in mCR2aa medium for 4 h in a CO$_2$ incubator (5% CO$_2$ in air, 90–95% relative humidity) at 38.5°C.

Chapter 11

Reproductive Cloning

Introduction

Selection criteria have been applied since the beginning of livestock rearing for propagation of animals with more desirable traits. With time, it was felt that the expansion of the desirable herds is limited either by the reproductive capacity of the species or breed, prevalence of particular versions of genes responsible for those traits in the available gene pool or limited female contribution to reproductive success. The female contribution to reproductive success is limited by species-specific characteristics such as average litter size, frequency of estrus and gestation length, while male contribution is restricted by degree of proximity to fertile females and capability to inseminate females with sufficient number of normal sperm. To overcome these limitations, various forms of assisted reproductive techniques have been implemented in livestock agriculture. These technologies form a continuum ranging from the fairly minimum assistance provided to animals engaged in natural service through those containing components of significant *in vitro* manipulation such as *in vitro* fertilization and embryo splitting, to the more recent development of somatic cell nuclear transfer (SCNT), colloquially referred to as cloning.

Reproductive cloning in general includes production of offspring by embryo splitting and nuclear transfer. Embryo splitting occurs naturally or can be induced artificially to form two or more genetically identical animals. Reproductive cloning by nuclear transfer refers to the creation of animals from a reconstructed embryo made by transferring the nucleus of a donor cell into an oocyte from which its genetic material has been removed. The donor cell could, theoretically, be any adult or fetal somatic cell of the body, embryonic stem cell, adult stem cell, mesenchymal stem cell as well as cells isolated from body fluids like urine, milk or semen.

Somatic cell nuclear transfer

Edward Dreisch (1885) was the first to demonstrate artificial embryo twinning in sea urchin. He showed that by merely shaking the two-celled sea urchin embryos, the cells could be separated and each cell could grow into a complete sea urchin. This experiment showed that each cell in the early embryo has its own complete set of genetic instructions and can grow into a full organism. With this began the interest in vertebrate embryo slitting where the first feat was achieved by Hans Spemann (1902) when he split the salamander embryo by tightening a tiny noose from a strand of baby hair between two cells of a salamander embryo.

Each of the two cells grew into an adult salamander. He also tried to divide more advanced salamander embryos by this method, but found that cells from these embryos were not as successful at developing into adult salamanders. Hence, it was concluded that cells from a more complex animal can be twinned to form multiple identical organisms but only up to a certain stage in development. This was followed by successful nuclear transfer by Briggs and King in frog. They transferred the nucleus from an early tadpole embryo into an enucleated frog egg, resulting into development of a tadpole. They also showed that cloning is less successful with donor nuclei from more advanced embryos. These experiments reinforced the conclusions drawn earlier that it is the nucleus that directs cell growth and development and the embryonic cells in early development are better for cloning than cells at later stages. The first experiment to prove that nuclei from somatic cells from a fully developed animal could be used for cloning was performed by John Gurdon (1958). He transplanted the nucleus from a tadpole intestinal cell into an enucleated frog egg and created tadpoles that were genetically identical (clones) to the one from which the intestinal cell was taken. Though Gurdon's nuclear transfer tadpole clones failed to metamorphose into frogs; the experiments, nevertheless, suggested that cells retain all of their genetic material even as they divide and differentiate. The success in amphibian cloning was taken with enmity and intrigue by the researchers working on mammalian eggs. The successful cloning in mammals demanded *in vitro* culture systems and micromanipulation equipments as well as better understanding of mammalian oocyte biology. The much smaller mammalian oocytes required finer tools and better pressure control. The culture systems which would allow continued development of embryos also needed to be developed. The limited number of eggs that could be obtained from a mammal was also a big obstacle. Despite of all these limitations, success in somatic cell nuclear transfer was achieved in diverse mammalian species, like rabbit, mouse, sheep, goat, pig, cattle, buffalo, etc. The first report of successful mammalian cloning was from Derek Bromhall (1975). He transferred the nucleus from a rabbit embryo cell into an enucleated rabbit egg cell and obtained morula after a couple of days. For a brief period of time (late 1980s and early 1990s), success in mammalian nuclear transfer was limited mainly to domestic animals. Steen Willadsen (1984), for example, used a chemical process to separate one cell from an 8-cell lamb embryo which he fused to an enucleated egg cell using a small electric shock. After a few days, he placed the embryos into the womb of surrogate mother sheep, resulting into birth of three live lambs. A somewhat similar procedure was followed by First, Prather and Eyestone for cloning in cow (1987). Using embryonic cells, they reported two cloned calves – Fusion and Copy. All these cloning experiments used embryonic blastomers, primarily from morula-stage embryos as nuclear donors. While multiple clones could be produced, the number of identical offspring were limited by the number of blastomeres per donor embryo as multiple rounds of cloning (re-cloning) resulted in decreased efficiency. Ian Wilmut and Keith Campbell (1996) first demonstrated the use of laboratory cultured cells as donor cells. They transferred the nuclei from these cells into enucleated sheep egg cells and produced two clones – Megan and Morag, thereby showing for the first time that cultured cells could be used as donor adult sheep' cells for cloning purposes. In another landmark experiments, Wilmut and Campbell created a lamb by transferring the nucleus from an adult sheep's udder cell into an enucleated egg. Of the 277 attempts that they made, only one produced an embryo that was carried to term in a surrogate mother, giving birth to famous lamb – Dolly. This experiment overturned the dogma in biology concerning the nuclear totipotency from adult cells and opened new opportunities and directions in research. This also led to interest in human cloning experiments to derive human embryonic stem cell lines. Since, primates are a good model for studying human disorders; it was felt that cloning identical primates would decrease the genetic variation in research animals and therefore, the number of animals needed in research studies. Similar to previous experiments, scientists fused early stage embryonic cells with enucleated monkey egg cells using a small electrical shock and implanted the resulting embryos into surrogate mothers. Two clones, Neti and Ditto, were produced in such experiments by Wolf's team (1997), providing evidence that human's closest relatives could be cloned. The further advance was made by production of cloned transgenic sheep, Polly, using transgenic sheep skin cells containing human Factor IX. Among all the species, mice were observed to be difficult to clone. This was attributed to early activation of its embryonic genome. Since then clones of pig (2000), goat (2002), swamp buffalo (2006) as well as some endangered animals have been produced. Our Laboratory at Animal Biotechnology center, National Dairy Research Institute, was the first to clone riverine buffalo (Murrah) in 2008. Till then we have been successful in producing a number of cloned embryos as well as live births, using a range of donor cells like embryonic stem cells, fetal and adult fibroblasts, semen epithelial cells, somatic cells isolated from milk and urine, etc. Overcoming the decades of technical challenges, Mitalipov and colleagues (2013) were the first to use somatic cell nuclear transfer to create human embryo that could be used as a source of embryonic stem cells.

Methodology of nuclear cloning

Nuclear cloning, in general, comprises a sequence of five main steps, as detailed below:

1. Maturation and enucleation of oocytes

Oocytes, commonly obtained from aspiration of follicles of ovaries obtained from slaughtered animals (cattle and buffalo), are matured *in vitro*. This is followed by enucleation which involves the physical removal of metaphase chromosomes and extruded first polar body. This removes the nuclear genetic material of the oocyte, resulting in formation of cytoplast (a cell containing only cytoplasmic material). However, mitochondrial DNA remains present within cytoplasm of the enucleated oocyte.

A number of approaches have been employed for enucleation of a matured oocyte. Since the mammalian oocyte is very small and difficult to manipulate manually, it demands the development of a technique which would combine microscopy with tiny holding and cutting tools. A combination of such tiny glass tools connected to precise electric motor-driven robot arms which are in turn linked to a specialized microscope is called a micromanipulator. The tips of these micro-tools are sometimes even smaller than the diameter of a sperm head, and need special dampened hydraulic and pneumatic controls to limit the comparatively gross movements of even the most delicate human hand. The micro-tools can perform effective microsurgery as the eggs or embryos as well as the tips of these micro-tools are magnified 800 times as cell surgeon operates. The early nuclear transfer procedures were not based on micromanipulation. The first nuclear transfer procedure by Jacques Loeb in sea urchin occurred by accidental osmotic blebing of cytoplasm, while in another model Spemann used hair of his newborn son to make transient and complete separation with a noose in salamander embryos. Briggs and King (1952) also did not use micromanipulation either as the bigger size of the frog egg and lack of zona pellucida allowed application of hand pipettes. The technically most difficult part of their experiment was not oocyte enucleation but preparation of donor cell nucleus, as removal of cytoplasm is a prerequisite for appropriate reprogramming in frogs. It was eventually the start of mammalian cloning. More specifically, a common belief that zona pellucida is indispensable for early embryonic development connected nuclear transfer to micromanipulators, according to most of the scientists. In 1981 a micromanipulator based cloning technique (MBCT) has been described by Illimense and Hoppe who reported the birth of the first cloned mouse embryo into an enucleated mouse zygote. With slight modifications, the most widely used and popular MBCT is the one described by McGrath and Solter (1983). This technique was used for cloning in various animals like cattle, rabbit, sheep and pig. Over the past two decades, more than 99% of scientific publications dealing with somatic cell cloning referred to micromanipulation-based enucleation and nuclear transfer, either by subzonal or intracytoplasmic injection of the donor cell or nucleus, respectively. Consequently, nuclear transfer remained the privilege of selected laboratories that could afford the considerable investment regarding both instrumentation and skills. As a consequence, costs were high and financing of this type of research frequently required commercial contribution. Despite improvements and numerous advances in all the facets of MBCT, this technique needs expensive equipment, certain laboratory arrangements and a considerable laboratory skill which limits its widespread application. Also almost all the scientists or groups have developed their own special version of micromanipulation making standardization of nuclear transfer and comparison of results between laboratories rather difficult. Micromanipulation, being a sophisticated procedure and dependent on many factors that can only be controlled by exceptionally well-trained scientists with considerable experience in the field, adds more inconsistencies to the nuclear transfer procedure.

The first to give a hope in these difficulties were Taniguchi et. al. (1992) who utilized a manual oocyte dissection described by Tarkowiski and Rossant (1976) to dissect mouse oocytes into karyoplasts and cytoplasts. They electrofused the cytoplasts obtained, onto mouse late 2-cell stage blastomeres to produce reconstituted embryos. In order to make the dissection simple, easy and rapid several modifications were done to this technique. One such modification involved deformation of zona-free eggs into cylindrical rods by sucking them into a glass pipette with a narrow mouth. Then the rods are dissected by a fine glass needle (5 μm) on the surface of agar under stereomicroscope. Another modification known popularly as Hand-Made cloning technique, was dissection of oocytes into karyoplasts and cytoplasts under stereomicroscope with Ultrasharp Splitting Blades. An almost similar procedure was followed in our laboratory for dissection of buffalo oocytes. We used fine Microblades for dissection of the oocyte under a zoom stereomicroscope into two parts; one part of the oocyte containing the genetic material (polar body being visible there) and another part only ooplasm. This technique has been rechristened as Hand-Guided cloning technique. In our lab, we employ dissection of *in vitro* matured buffalo oocytes following visualization of polar body at one end of matured oocyte. The matured cumulus oocyte complexes (COCs) with expanded cumulus are treated with Hyaluronidase for 2–10 min and followed by vortexing for 2 min to remove the cumulus cells. The completely denuded oocytes with evenly granular cytoplasm are incubated in Pronase for complete zona pellucida digestion. This is followed by incubation for almost 30 min, so that the protrusion cone becomes prominent and is easily visible. The protrusion cone bearing oocytes are transferred (5–10 each time) into 35 mm dish for treatment with Cytochalasin B. The treated oocytes are then fixed under microscope and bisected using a Microblade in such a way that the protrusion cone remains with the smaller half. After bisection the larger demicytoplasts without protrusion cone are incubated so that they regain their spherical shape.

A non invasive chemical enucleation procedure was also employed in mouse oocytes by Fluka and co-workers (2001). This basic idea of this procedure is to block DNA topoisomerase II enzyme (top II) during metaphase I (M I). This inhibits oocyte chromosome separation and the whole chromatin is expelled into the first polar body leaving a chromatin-free cytoplast, the so called chemically enucleated oocyte (CEO). This procedure has been employed by Elsheikh and his colleagues (1997) for cloning of mouse embryos by fusion of CEO to late 2-cell stage mouse embryos.

2. Injection of donor DNA

This step involves injection of single donor cell underneath outer zona pellucida and adjacent to the cytoplast membrane. The donor cell can also be attached to the membrane of a zona free cytoplast (zona free cloning) by a chemical treatment. The donor cells can be from a variety of sources representing different degrees of cellular differentiation. For example, the donor cells could be embryonic blastomeres, cell lines such as embryonic stem cells, or primary cultures from biopsies obtained from selected bulls. In fact all kinds of nuclei from early embryonic stage such as pronuclei from zygote, embryonic blastomeres, inner cell mass of the embryo, embryonic stem cells, cumulus cells or any other somatic cell can be used as nuclear donor in cloning, only if they can be isolated without destroying them. In case of buffalo zona free cloning, the enucleated demicytoplasts, as obtained above, are immersed in Phytohemagglutinin (0.5mg/ ml) for 3–4 sec and transferred into a medium containing low density donor cells. Each demicytoplast is then allowed to attach to a single, rounded, medium sized cell by gently rolling the demicytoplast over it. The couplets (demicytoplast –donor cell pairs) are transferred to fusion medium for equilibration. The couplets and remaining demicytoplasts are then transferred away from positive and negative poles, respectively, of the fusion chamber of the BTX microslide. This is followed by a single step fusion protocol wherein a demicytoplast and a couplet are picked using a fine pulled capillary pipette (Unopette). The couplet is initially aligned with an AC pulse (4 volt) using BTX Electrocell Manipulator 200, so that the donor cell faces the negative electrode. After this a demicytoplast is introduced into the fusion chamber close to the somatic cell, so as to sandwich the somatic cell in between the cytoplasts. A single DC pulse (3.36kV/ cm for 4 μsec) is then applied to fuse donor cells with the two cytoplasts (triplets). The triplets are then incubated in an appropriate medium for rounding up and complete fusion for up to 4 h at 38.5°C.

3. Fusion of cytoplast and donor cell

The cytoplast and donor cell are fused together utilizing an electrical field, as described above for buffalo cloning. Thus the genetic information contained within the nucleus of the donor cell enters the cytoplast. This is the essence of the term nuclear transfer, whereby the genetic information from the oocyte is removed and is replaced with that from donor cell. After reconstruction the donor nucleus has the opportunity to be reprogrammed following molecular interactions between factors present in the oocyte cytoplasm and the donor chromatin.

4. Activation of the reconstructs

The reconstructed 1-cell embryos are artificially activated using either specific chemical signals or electrical pulses, in order to initiate embryonic development. For bubaline species, the reconstructed oocytes are activated by incubation in a medium containing 5 μM Calcimycin A23187 for 5 min at 38.5°C followed by incubation in the culture medium containing 2 mM 6-Dimethylaminopurine (6-DMAP) covered with mineral oil in a CO_2 incubator at 38.5°C for 4 h. The activation procedure should be carried within 29 h after start of maturation. The mechanism of activation is similar to parthenognetic activation, as discussed in the previous chapter.

5. Culture of the embryos

Following activation, the reconstructed embryos are cultured *in vitro* in a chemically-defined medium for a specific time period (7–8 days for cattle and buffalo). After this time, embryos that have developed into blastocysts of suitable quality are used for further studies, like nuclear transfer to produce cloned calves or development of nuclear transfer (NT) embryonic stem cell lines, or some other purposes. Bubaline embryos are successfully cultured in modified Charles Rosenkrans 2 medium supplemented with amino acids (mCR2aa) and 1% fatty acid free bovine serum albumin or in modified synthetic oviductal fluid with essential and nonessential amino acids, sodium citrate and myoinositol (mSOFaaci) with 1% fatty acid free bovine serum albumin or in Research Vitro Cleavage medium (RVCL) supplemented with 1% fatty acid free bovine serum albumin. The reconstructed embryos are usually cultured (10–15 embryos per well) in 400 µL of respective media on a flat surface in each well of a 4-well dish, covered with 400 µL mineral oil and kept undisturbed in a CO_2 incubator for 7 days.

Factors affecting cloning

Cloning in farm animals involves variable raw material, work with extremely sensitive cells and systems and use of many chemicals and tools of inconsistent and uncomfortable quality. Though all experts in the field use good techniques and highly structured solution to avoid most inconsistencies generated by these factors, the solutions, even if completely acceptable in a given situation, differ from laboratory to laboratory or from person to person. The ever present human factor also adds to inconsistency. Even the day of the week play a role in a sense that the work performed during second half of the week may be more efficient than of the first half as the early and more sensitive period of subsequent embryo development falls on weekend which is relatively free of incubator disturbances. Some of the predominant biological and technical factors known to affect cloning efficiency are dealt with in this section.

1. Oocyte source and quality

Oocytes are usually collected from slaughterhouse-derived ovaries or live cows by ovum pick-up (OPU) and used for *in vitro* maturation. Most of the studies have suggested that inappropriate oocyte maturation of recipient cytoplasts as an important factor causing embryonic or fetal loss after nuclear transfer in almost all the domestic animal cloning experiments. Thus, the factors affecting oocyte maturation *in vitro* should be considered functional for SCNT efficiency. The readers are advised to refer to the chapter on *in vitro* maturation for the detailed discussion regarding these factors. In general, good quality (A + B grade COCs) show better cloning efficiency than poor quality oocytes. It has also been observed in cattle cloning that pre-treatment of OPU-derived COCs with follicle stimulating hormone leads to significant increase in cloning efficiency in comparison to non-treated OPU COCs and aspirated COCs from slaughterhouse derived COCs. The blastocyst formation rate of *in vivo* matured oocytes collected by OPU from heifers treated with FSH, PGF2α and GnRH was found to be higher in comparison to *in vitro* matured COCs. The pregnancy rate, however, did not differ between *in vivo* and *in vitro* matured oocytes, though higher abortion rate was observed in SCNT fetuses from *in vitro* matured oocytes.

2. Technical challenges

Animal cloning involves a manual manipulation process *in vitro* requiring exposure of oocytes and cells to fluctuations in temperature, atmospheric conditions, light, media variations, *p*H changes, etc. The exposure of oocytes to physical and chemical insults, through procedures for enucleation, like micromanipulation, chemicals, zona removal, etc., as well as transfer and fusion of donor nuclei and activation stimulus, affect the overall oocyte health and developmental competence. A minor alteration at any one step in the process affects the success rate significantly. Thus, species-specific requirements and sensitivities need to be identified to achieve successful nuclear transfer.

3. Donor cell

Cloned calves have been produced from various somatic cell types. However, it is still unclear which cell type is the most appropriate for bovine or buffalo cloning. The source of donor nucleus, its epigenetic status, duration of culture, differentiation status and stage of cell cycle are among the various known factors which affect cloning efficiency.

The discussion on the source of the donor cell has ranged from initial disbelief that it is possible to clone using adult cells to theories that the low success rate is contributable to the low percentage of stem cells that may be present in adult tissues and inadvertently used as donor cells. While cells from various adult tissues had been used to produce offspring in sheep, cattle and mice, the concept that a truly differentiated cell could be reprogrammed by SCNT was proven when mature mouse B and T cells were used as donor cells. It has been demonstrated that nuclei containing chromosomes that had undergone rearrangement, a definitive step in cell differentiation, could result in live offspring following SCNT, albeit at very low rates and in some cases after resorting to chimera production to achieve success. Some researchers have demonstrated that differentiation status of somatic cells has no relationship with cloning efficiency. It has been proved that bovine SCNT embryos develop to blastocyst stage at a rate similar to that of embryos produced by IVF, although electric conditions for fusion of enucleated oocytes differ among donor cell types. However, high embryonic and fetal looses occur after embryo transfer regardless of donor cell type. A comparison between cumulus, fibroblast and mammary epithelial donor cells used in cattle cloning, although, revealed no differences in cleavage rates of embryos, but it were cumulus donor cells that resulted to highest rate of blastocyst development, while the poorest rate of *in vitro* development and no full term survival was seen in mammery epithelial cells. The donor age has also been found to affect cloning efficiency. It has been concluded that cells from fetuses and newborn animals are more efficient in nuclear transfer. However, little changes are observed in cloning efficiency when cells from cattle varying from 2 to 16 years of age were used as donor cells. It has been observed that clones derived from adult cells frequently abort in later stages of pregnancy and calves which developed to full term show a higher number of abnormalities than those derived from new born or fetal cells. It appears from the large number of studies conducted in cattle and buffalo cloning, that cells from fetuses as well as aged adults can lead to comparable blastocyst development of cloned embryos. However, the fetal cells, in general, have been found to be better than adult cells in producing healthy live births. This might be due to the fact that somatic cells of adult animals have accumulated more genetic mutations, are more terminally differentiated than fetal cells and are thus more likely to fall at full term development. The cell culture duration of donor cells also has its implications in cloning efficiency. It has been found in cattle that cells (fibroblasts) of later passages yield more cloning efficiency than those at lower passages. This demonstration that later passages can support clone development is essential for utilizing somatic cloning for gene-knockout studies, in which single cells must be clonally expanded to generate sufficient cells for nuclear transfer. It has been suggested that cells of higher passages were receptive to nuclear reprogramming and contain lesser epigenetic modifications (histones are more acetylated than in earlier passages). It has also been demonstrated that *in vitro* culture of cells can induce expression of genes that are not expressed before culture and a greater proportion of late passage cells (passage 18) are found to be in G0/G1 cell cycle state than earlier passage cells (passage 2), whether or not they are in serum-starved culture conditions. It has also been demonstrated that coordination of cell cycles between donor karyoplast and recipient cytoplast is critical for efficient cloning process. However, the coordination of cell cycles presents technical and biological challenges. The maintenance of donor cells and recipient oocytes at a particular point in their cell cycle for a particular time required for the cloning process detrimentally affects their viability. The final aim of cell cycle coordination is to bring donor cells in G0/G1 state. Some of the methods employed to arrest donor cell cycle are serum starvation, exposure to pharmaceuticals like roscovitine, nocodazol or non-pharmaceutical treatments like mitotic shake-off to select recently divided cells in G1 or contact inhibition to select cells in G0/G1. However, these methods are harmful to donor cell health, for example, serum starvation can adversely affect chromosome integrity and the exposure to chemicals also has toxic side effects on the donor cells.

4. Fusion and activation

In SCNT, the lack of sperm-induced fertilization necessitates artificial activation to trigger further development. The activation is performed by application of an artificial stimulus (physical or chemical) to produce a brief increase in calcium which is usually followed by inhibition of phosphorylation or protein synthesis using 6-dimethylaminopurine (DMAP) or cycloheximide, respectively. This treatment results in decrease in maturation promoting factor and mitogen activated protein kinase, which allows the reconstructed zygote to form a pronucleus and start the developmental process. The timing of activation of MII oocytes can be classified into two protocols: i) activation performed *immediately* after fusion (simultaneous fusion and activation method, FA) and ii) activation performed several hours after fusion (*delayed* activation method, DA). Both FA and DA methods have been successfully employed to produce cloned bovine offspring. Donor chromosomes are exposed to factors present in MII cytoplasm for only a short time in FA method and for a longer time in DA method. Since, the direct exposure of chromosomes to nonactivated MII cytoplasm is effective for somatic cell nuclear reprogramming; the DA method improves *in vitro* development of embryos derived from somatic cells at G0/G1 stage compared with that of the FA method.

We employ the DA method for buffalo cloning in our laboratory, performing fusion at 21 h of maturation and activation 4 h later. It has been found in bovine cloning that development of embryos activated 6 h after fusion in the DA method is significantly lower than that in the FA method. The excessive exposure to MII cytoplasm has been found to result in abnormal chromatin morphology and thus reduced developmental competence. The SCNT embryos activated less than 2.5 h after fusion has been shown to result in improved nuclear morphology and increased development to the compacted morula/ blastocyst stage.

Epigenetic reprogramming of donor nucleus

For normal embryogenesis to occur after nuclear transfer, it is generally accepted that the donor nuclei must be reprogrammed to a state comparable to that in a zygote to allow for the correct pattern of subsequent gene expression. This reprogramming must occur in a short time frame and in a different cellular context compared with normal development and is thus prone to errors. SCNT should also erase the cell-type specific memory that has been imposed during differentiation. Thus, SCNT should somehow overcome these two epigenetic hurdles of somatic cell marking and cell-type specific differentiation memory. Each hurdle might cause specific reprogramming errors and clone-associated abnormalities. It is being presumed that the alterations in gene expression after nuclear transfer are caused by epigenetic errors primarily in the patterns of DNA methylation and chromatin organization. One of the most important regulators of gene expression is chromatin architecture which manifests itself in DNA methylation, nuclear lamin composition, histone subunits and subsequent post-translational modifications such as acetylation, phosphorylation, methylation, etc. For SCNT to be successful, it is thought that the donor nucleus must be remodeled to resemble the nucleus of a zygote. The cytoplasm of an oocyte arrested in metaphase can facilitate that remodeling. This change in chromatin architecture (nuclear remodeling) results in a change in pattern of genes that undergo transcription, resulting to nuclear reprogramming. Thus, SCNT results to both nuclear remodeling (physical repackaging of DNA) as well as nuclear reprogramming (changes in gene expression levels).

Nuclear remodeling during cloning

The structure and functions of the pronuclei in a zygote are rather unique in that they are organized in such a way that in the environment of the zygote cytoplasm there is very little, if any, transcription occurring. The first few cleavage divisions are directed by factors stored in the cytoplasm of the oocyte. Depending on the species, the embryo begins to produce significant amounts of RNA at a certain cleavage stage. The point at which significant amounts of transcription occur is called *embryonic genome activation* (EGA). It occurs at 2-cell stage in mouse, 4-cell stage in human, rat and pig, and during 8–16 cell stage in sheep and cow, while in *Xenopus* it occurs at 12th cleavage division (~4000 cells) around the time of mid-blastula transition. At this point, the nuclei in the developing embryo begin truly controlling the development of the embryo. Upon initiation of EGA, the proteins associated with nucleus, like nuclear lamins, SnRNPs, and histones change their associated acetylation and methylation patterns. During the differentiation of an embryo into distinct inner cell mass and trophectoderm, a different set of proteins are associated with the nuclei (CDX2 and POU5F1) in addition to different set of genes that undergo transcription. As the development proceeds and various tissues form and begin to specialize, each tissue type has its specific nuclear structure and repertoire of genes that are turned on. Thus, when a somatic or any other nucleus is transferred into metaphase arrested ooplasm, it is poised to recapitulate the same pattern of development observed in normally fertilized embryo. Thus, structural remodeling of the chromatin should result in subsequent reprogramming of the developmental pattern of gene expression that resembles the fertilized embryo. Another well characterized exchange of proteins involves histone H1 which links the nucleosomes together. A variant of H1 called H1FOO occurs in oocyte. When somatic cells are transferred to ooplam, the somatic H1 is rapidly replaced with the oocyte variant. Another histone variant present in somatic cells but absent in oocyte, like MacroH2A is removed from the somatic chromatin and degraded during the first few cleavage divisions. It is then re-synthesized and assembled into the chromatin structure when the embryo reaches the morula stage, coinciding with the normal appearance of MacroH2A in fertilized embryos. Methylation of DNA and modifications of histones also occur which vary from species to species. The global DNA methylation occurs during preimplantation development in mouse and is facilitated by active demethylation of paternal genome and passive demethylation of the maternal genome during cleavage, in most of the species studied. *De novo* DNA methylation begins around the time of ICM and TE differentiation, followed by lineage-specific methylation as the cells become more and more differentiated. However, porcine embryos show *de novo* methylation as early as at blastocyst stage, while sheep embryos undergo limited demethylation after fertilization or remethylaion from the zygote to the blastocyst stage. This pattern is not always replicated in cloned embryos. For example, in cloned 1-cell murine embryos reduced methylation is seen as compared to zygotes, and the additional demethylation that occurs in fertilized zygotes does not occur in cloned ones. Also in cattle SCNT embryos, passive demethylation process appears to be defective and the pattern of DNA methylation in these early cloned embryos is more similar to those of differentiated cells. This aberrant DNA methylation continues beyond the blastocyst stage. This also includes some imprinted genes such as SNRNP, H19 and IGF2. It has been concluded that if methylation pattern of the imprinted genes in donor cell is not set then it becomes very difficult for SCNT embryos to reestablish these absent imprints. A prime example is when the parental genomic imprint is erased during development of germ cells and after it is maternally or paternally reset, the clones from these donor cells do not develop. In germ cells where the parental imprints have been erased and in the mature germ cell where only the maternal or paternal imprint is present, SCNT embryo is not capable of establishing those imprints. Thus, beginning with the correct maternal and paternal imprint is necessary for development of the SCNT embryo. This incomplete remodeling of methylation pattern likely contributes to low efficiency of development in cloned embryos. Thus, the changes in the structure of the chromatin caused by an exchange of proteins, changes in DNA methylation and histone post-translational modifications (remodeling) results in different pattern of gene expression. This change in gene expression pattern results in a zygote pattern followed by a recapitulation of the normal developmental pattern (reprogramming).

Nuclear reprogramming during SCNT

Chromatin remodeling, as discussed, results in the recapitulation of the expression of many genes. Some of the first such genes studied in *Xenopus* are 5Sooc gene and muscle-specific actin, both of which serve to illustrate the remarkable fidelity of reprogramming. The 5Sooc gene is transcribed for only a short period of time during the late blastula stage. If a donor nucleus from an embryo beyond the blastula stage is transferred to an oocyte, the 5Sooc gene remains off until the late blastula stage when it is briefly transcribed and shut off. Similarly, muscle-specific actin is produced only in the developing myotome cells. If a nucleus is taken from a muscle cell and transferred to an oocyte then muscle specific actin is shut off. When the resulting embryo reaches the stage at which the myotome cells are differentiating, the muscle-specific actin turns on again, but only in the developing myotome cells of the embryo. In this context, many studies have evaluated gene expression in cloned mammalian embryos and tissues and it has been concluded that each set of donor cells appears to be different as some can be easily reprogrammed and others not. Though majority of genes appear to be correctly reprogrammed, there is a different subset of genes in many donor cells that are not reprogrammed and are expressed at inappropriate times. Thus, a consistent pattern of aberrant gene expression has not been identified, suggesting numerous flaws in genomic architecture after SCNT. For example, there is mounting evidence of more variation in DNA methylation within clones than between non-clonal controls and this also corresponds to phenotypic variation. It is presumed that if remodeling is not sufficient during the first cell cycle, all descendants would be affected.

Nonnuclear remodeling

In addition to the exchange of proteins between the transferred nucleus and the cytoplasm to restructure the chromatin, there must also be a synchronization of nonnuclear components of the cell such as the machinery required for cell division and mitochondria. For cell division, the chromatin must be configured such that the chromosomes can segregate appropriately during the first mitosis in which components of the spindle, like centrosomes and γ-tubulin play important role. In SCNT embryos, γ-tubulin accumulates at the metaphase spindle poles and is associated with abnormal chromosome segregation at first mitosis. The nuclear mitotic apparatus and other markers of spindles and chromosome segregation have been observed to be aberrantly distributed in SCNT embryos.

Anomalies in cloned animals

There is evidence that some cloned animals are physiologically normal or at least may develop a stable metabolism some time after birth. It is remarkable that SCNT is successful, for a tremendous amount is asked of a differentiated donor nucleus to re-establish the correct pattern of gene expression to allow normal embryogenesis. It has been proved in various studies that some clones appear to be same as their non-clone counterparts in behavior, growth rates, reproduction, livestock production characteristics and life spans. The sexually derived offspring of clones have also been shown to be normal. There are, however, various instances which show that reprogramming appears to be incomplete. As discussed earlier that for normal development, epigenetic modifications in the donor nucleus must be reprogrammed to a state comparable to that of zygote. This reprogramming must occur within short time frame, in a different cellular context compared with normal development and is prone to error. A large amount of data documents deviations in epigenetic reprogramming, with clones showing inappropriate patterns of DNA methylation, chromosome modification, X-chromosome inactivation, expression of imprinted and non-imprinted genes, placental abnormalities etc. A wide spectrum of phenotypic outcome, ranging from lethal to neutral, has been observed in cloned animals. The neutral outcomes do not compromise the health and welfare of animal but the epigenetic variations reduce the uniformity in the clonal family which may be undesirable for some applications. Some of the common consequences of incomplete reprogramming are higher rates of pregnancy loss, difficult parturition, higher rates of post-natal mortality and some specific clone-associated phenotypes in adulthood. We will discuss some of these prominent problems in this section.

1. Placental abnormalities

The failure of placenta to develop and function correctly is a common feature amongst the clones. The majority of early pregnancy failures before placentome formation are attributed to an inadequate transition from yolk sac to allantoic derived nutrition. There is also reported evidence of immunological rejection contributing to early embryonic loss. For example, in cattle 50% to 70% of pregnancies at day 50 are lost throughout the reminder of gestation and up to term, in stark contrast to only 0% to 5% with artificial insemination or natural mating over the same period. It has been observed that cloned placentae have only half the normal number of placentomes which display compensatory overgrowth and are oedematous. Another factor for gestational loss is hydroallantois, excess accumulation of fluid within the allantois. Its incidence in bovine clones has been reported to be 25% at 120 day of gestation, as compared to 0.07% to 5% for artificial fertilization and IVEP, respectively. The incidence of hydroallantois has also been found to vary with the individual cell line used as donor nucleus.

2. Parturition difficulties

The gestation length in SCNT pregnancies is typically prolonged with birth weight of cloned calves 25% heavier than normal, thus, necessitating the intervention to deliver the cloned offspring. It has been observed that newborn cloned calves display functional adrenal glands. The extended gestation may be due to failure of the placentae to respond to fetal cortisol near term or to lack of adrenocorticotropic hormone release from the fetus. Cloned calves are larger than IVEP, artificially inseminated and naturally mated controls. It has been reported that somatic cloned claves are heavier than embryonic clones.

3. Post natal viability

The viability of cloned offspring at delivery and up to weaning is reduced compared to normal, despite greater than usual veterinary care. It has been shown in some studies that 80% of cloned calves delivered at term are alive after 24 h. Two-thirds of the mortality within this period is due to spinal fracture syndrome through the cranial epiphyseal plate of the first lumbar vertebrae or to deaths that occur either *in utero* or from dystocia. The surviving newborn clones have altered neonatal metabolism and physiology, possibly due to placental abnormalities. It takes time for these processes to adjust to normal. Gastroenteritis and umbilical infections are among the most common mortality factors during the weaning period. Other abnormalities include defects in cardiovascular, musculoskeletal and neurological systems as well as susceptibility to lung infections and digestive disorders. It has been found that the proportion of cloned calves that survive weaning is significantly greater for those derived from quiescent G0 donor cells than those derived from G1 cells. Post natal survival of cloned sheep has been reported to be lesser than cattle with both somatic and embryonic cell types.

4. Clone-specific phenotypes

Enough evidences have been reported of abnormal clone-associated phenotypes that become apparent during the juvenile and adult phases of life. Their incidence varies according to the particular species, genotype or cell type or according to species aspects of nuclear transfer and culture protocols used. Some of these are thought to be effected by nuclear-mitochondrial interactions arising from a donor nucleus in a foreign cytoplasmic environment, effect of mitochondrial DNA heteroplasmy and possible recombination events. It is being presumed that the cloned offspring syndrome is a continuum in that the lethality or abnormal phenotypes may occur at any phase of development, depending on the degree of dysregulation of key genes, presumably due to fundamental errors in epigenetic reprogramming.

5. Trans-generational effects

Offspring of male and female clones in a range of species have been produced following both natural mating and assisted sexual reproduction, such as artificial insemination with a non-cloned partner. Conception, pregnancy, parturition and survival all have been found to be within normal ranges, as is the subsequent fertility of these offspring of clones. With more discriminatory mating of cloned males with cloned females performed in sheep, cattle and mice, no evidence of the placental abnormalities and large birth weights were recorded. Our observations in our own laboratory were also similar with Garima (cloned female buffalo using embryonic stem cell as the donor cell) when she gave birth to Mahima (female calf) upon artificial insemination. In our case we observed normal gestation time, normal birth weight of the offspring, and no placental or other abnormalities. Mahima is currently keeping sound physiological and reproductive health and we expect the second generation offspring from her through AI.

6. Abnormal X-chromosome inactivation

Mammals have evolved a sex determination mechanism that is chromosome based with the primary sex determinant for females being the presence of two X chromosomes, and the presence of an X and a Y chromosome for males. X chromosome consists of approximately 160Mb DNA that codes for over a thousand genes with diverse range of functions, while the Y chromosome is comparatively gene-depleted consisting of a variable amount of DNA that codes for less than 100 genes mainly involved in sex determination and fertility. To overcome the potential unequal expression of genes resulting from unequal copy number of chromosomes, female animals have developed an epigenetically regulated process of dosage compensation, known as X-chromosome inactivation. In general transcription of X-linked genes is restricted to a single, active X chromosome (Xa) while it is inhibited on the other inactive X chromosome (Xi). This X-inactivation process is initiated early in embryogenesis by transcription of XIST from one of the two X chromosomes that is to be inactivated and subsequent coating of the same X chromosome by the untranslated XIST RNA. The choice of which X chromosome becomes inactive appears to be under an imprinted control, where random X-inactivation occurs in the inner cell mass derivatives and preferential inactivation of the paternal X occurs in the trophoblast derivatives. Immediately after XIST RNA coating begins, the Xi undergoes various chromatin modifications such as loss of methylation on H3 lysine 4 (H3-K4), methylation on histone H3 lusine 9 (H3K9) and methylation on H3 lysine 27 (H3K27), leading to transcriptional silencing and late replication of Xi. Other chromatin modifications include hypoacetylation of histone H4, macroH2A recruitment and DNA methylation. The functional links between methylated DNA and histones are extremely stable and are maintained throughout all subsequent cell divisions and life. In mouse, human and bovine embryos, XIST coating occurs at two-, four-, and eight cell stages, respectively. The relative expression levels of a small panel of X-linked genes in bovine male and female morulae and blastocysts showed that the process of dosage compensation or equalization of expression of genes like PGK, G6PD and HPRT is initiated or completed in the blastocyst stage embryos. Thus, it appears that X-inactivation is initiated and established as the embryo develops from early cleavage stages to the blastocyst stage. It has been observed that rate of development to the blastocyst stage differs between males and females; with males developing more rapidly. This skews the sex ratio towards males under IVEP. The effect of medium has also been observed on sex ratio; with medium rich in glucose yielding a preponderance of male blastocysts on day 7 after *in vitro* fertilization. Culture media supplemented with serum and factors from co-cultured somatic cells tend to have preponderance of female embryos arrested at the morula stage. The exposure of IVP embryos to elevated temperatures has also been shown to cause a greater loss of female embryos. In case of SCNT, faulty X-chromosome inactivation counts among the main epigenetic factors regulating the development of embryos. Some of the embryos and offspring have been shown to exhibit aberrations in X-chromosome inactivation. Although the inactive X of the donor cells can be successfully reactivated by the recipient cytoplast, heterogeneticity within SCNT mouse blastocysts for X-inactivation has been observed, with cells showing 0, 1 and 2 inactive X chromosomes. The kinetics of preferential paternal X inactivation differed significantly between normal and cloned mouse embryos, the latter showing abnormal XIST expression pattern. In bovine embryos, XIST RNA, the initiator of X-inactivation, was found in samples taken from pools of male SCNT blastocysts, but not in male IVP or *in vivo* blastocysts. The pattern of X-inactivation in aborted bovine SCNT fetuses and dead newborn calves was found to be altered. The placental samples exhibited random X-inactivation as opposed to non-random preferential paternal X-inactivation seen in normal controls and healthy SCNT calves. It has been proved beyond doubt that faulty removal of existing epigenetic marks and subsequent reprogramming of the donor nucleus during SCNT leads to abnormal patterns of X-chromosome inactivation that is deleterious for development of female SCNT embryos. To remedy the faulty reprogramming of X chromosomes after SCNT, DNA demethylating agents such as 5-aza-2'-deoxycytidine and S-adenosyl homocysteine have been used on donor cells to chemically reactivate the Xi prior to nuclear transfer. Both approaches have resulted in improvement in the rate of development to the blastocyst stage.

6. Telomere length and SCNT

Telomeres are the natural ends of linear chromosomes that play a critical role in maintaining the integrity of chromosomal DNA by preventing loss of terminal coding sequences and preventing end-to-end chromosome fusion. These are composed of repetitive DNA elements and specific DNA proteins, which together form a nucleoprotein complex at the ends of eukaryotic chromosomes. A specialized RNA-dependent DNA polymerase, the telomerase, is required to maintain the natural length of telomeric DNA. This enzyme is active in hematopoietic, cancer and germ cells and in early embryos at the blastocyst stage. As a general rule, some loss of telomere length occurs with each cell division as a result of incomplete replication of lagging strand. The telomeres of Dolly were found shorter in comparison to age-matched naturally bred counterparts, but correlated with telomere length of the donor cells. However, the vast majority of cloning studies reported the telomere length in cloned cattle, pigs, goats and mice are comparable with age-matched, naturally bred controls even when senescent donor cells were used for cloning. The regulation of telomere length is to some extent related to the type of donor cells employed for cloning. It has been found that telomere length in cloned cattle from fibroblasts or muscle cells is similar to that of age-matched controls, while clones derived from epithelial cells do not have telomeres of normal length. A check point for the elongation of telomeres to their species determined length has been discovered at the morula to blastocyst transition in bovine and mouse embryos. This morula/blastocyst transition is a critical step in the preimplantation development leading to first differentiation into two cell lineages, inner cell mass and trophoblast, which coincides with dramatic changes in morphology and gene expression. The telomeres are at the level of the donor cells in SCNT morulae, wheres at the blastocyst stage they are resorted to normal length.

Improving nuclear remodeling and reprogramming

It is evident from the foregoing discussions that remodeling and reprogramming in clones is not complete. Since, reprogramming depends on remodeling, the basic question is to devise methodologies to improve remodeling which would bear a direct effect on reprogramming. A number of methods have been attempted to facilitate sufficient remodeling during SCNT. Some of these methods are a subject of this section.

1. Protein exchange

The exchange of proteins between nucleus and cytoplasm, as discussed earlier, is one of the main things that occur during remodeling. This exchange probably needs to be at the level of DNA. Since chromatin is packaged and there are physical constraints on the ability of proteins to exchange with chromatin, any treatment that would open up the chromatin to make it more accessible may improve nuclear remodeling. It becomes necessary to remind that the three-dimensional structure of the nucleus prevents a given gene from undergoing both transcription and replication at the same time, as both DNA polymerase and RNA polymerase cannot occupy the same physical location at the same time. Transcription generally occurs during G1 and G2 phases of the cell cycle, while DNA synthesis occurs during S phase. As the early embryos have very short, if not nonexistent, G1 and G2 phases, the proteins responsible for remodeling are competing with a very tight packaging that is responsible for chromosome condensation during mitosis, and if additional remodeling occurs during interphase these proteins would be competing with DNA polymerase during S phase of the first cell cycle. It has been shown that fusion of cells together for creation of hybrids or simply addition of a transcription factor (s) results in dramatic change in transcription. These methods are generally known as transdifferentiation. A similar approach by using cell free system has resulted in the uptake and assembly of T-cell specific factors into fibroblast nuclei. Similar cell free systems may aid in elucidating the mechanisms responsible for the remodeling properties found in oocyte cytoplasm.

2. Inhibition of histone deacetylases

The structure of transferred nucleus is not always modified by oocyte cytoplasm, thus, necessitating some treatments that could help facilitate the change. One of such commonly used methods is inhibition of histone deacetylases. Trichostatin A (TSA) and 6-(1, 3-dioxo-1H,

3H-benzo (deisoquinolin-2-yl)-hexanoic acid hydroxyamide (Scriptaid) are ell known potent histone deacetylase inhibitors (HDACi). Inhibition of these deacetylases results in an increase in the global acetylation of histones. This consequently results in increased acetylation which further results in change in chromatin structure such that proteins like RNA polymerases can gain access to DNA and begin transcription. It has been reported that TSA or sodium butyrate treatment improves development of cloned embryos to the blastocyst stage. In mice, TSA treatment has been found to improve both the nuclear remodeling and development to term. However, other groups have reported neonatal death after TSA treatment of rabbit and pig embryos. In buffalo TSA results reported have also been controversial. TSA treatment has also been shown to result in severity of placentomegaly. Treating reconstructed pig zygotes, not donor cells, with Scriptaid results in an increase in histone acetylation intensity in 1-cell stage SCNT embryo to a level that is similar to IVF embryos at the same stage.

2. Altering DNA methylation

DNA methylation changes also change three-dimensional structure of chromatin to provide for nuclear remodeling. One of the most widely used chemicals that alter DNA methylation is 5-aza-20-deoxycytidine (5-aza-dC). It reduces DNA methylation in donor cells. The treatmen of donor cells with 5-aza-dC has been found not to result in an increase in development *in vitro* or *in vivo* in bovines. However, treatment of bovine donor cells and embryos with both TSA and 5-aza-dC results to increased histone acetylation, decreased DNA methylation and improved blastocyst development.

3. Inhibition of proteasomal machinery

Inhibition of proteasomes is thought to result in maintenance of recipient oocyte in metaphase by preventing the degradation of Cyclin B. It has been shown that treatment of rat oocytes, flushed from oviduct, with specific and reversible proteasomal inhibitor, MG132, prevents their precocious activation by prevention of Cyclin B degradation. Proteasomal inhibition also facilitates nuclear remodeling by preventing degradation of such factors present in ooplasm which promote nuclear remodeling.

4. Chromosome transfer

It is being presumed that removal of as many proteins associated with the donor nucleus as possible prior to nuclear transfer, so that only the factors affecting nuclear architecture of the clone are from ooplasm would increase cloning efficiency. Thus, it might be beneficial to transfer chromosomes in metaphase or telophase rather than in interphase for enhancing cloning efficiency. This should result in packaging of SCNT nucleus with as few factors from the donor cell as possible and as many factors from the ooplasm as possible. However, with these procedures cloning efficiency has not improved greatly and cloning abnormalities have also been encountered as well.

Applications of SCNT

SCNT has the potential to impact animal breeding in as fundamental a manner as artificial insemination. Given its current high costs (about $20,000 for a live calf) and relatively low success rates (<10%), SCNT will likely be used to improve production characteristics of food producing animals by providing breeding animals and not the production animals. Cloning has the relative advantage over other assisted reproductive techniques, discussed earlier, of allowing for the propagation of animals with known phenotypes to serve as additional breeding animals. This is critically important in breeding programs, especially when it may take years to prove the merit of a sire or dam. However, a number of issues are to be addressed before commercial opportunities for cloning in livestock agriculture can be realized. Some applications of cloning technology for agriculture and medicine are briefly discussed in this section.

1. Rapid multiplication of desired livestock

Cloning could enable the rapid dissemination of superior genotypes from nucleus breeding flocks and herds, directly to commercial farmers. It can provide genotypes ideally suited for specific product characteristics, disease resistance or environmental conditions. Cloning could be extremely useful in multiplying outstanding F1 crossbred animals or composite breeds to maximize the benefits of both heterosis and potential uniformity within the clonal family. Given that cloning is not particularly efficient at present, a niche opportunity exists in the production of small numbers of cloned animals with superior genetics for breeding. This could be relevant to mutton and beef industries, where cloned sires could be used in widespread natural mating to provide an effective means of disseminating their superior genetics. This could be used as a substitute for artificial insemination, which in these more extensive industries is often expensive and inconvenient.

2. Animal conservation

Cloning together with other forms of assisted reproduction could be used to preserve indigenous breeds of livestock, which have production traits and adaptability to local environments that should not be lost from the global gene pool. Interspecies nuclear transfer and embryo transfer may also be used to aid the conservation of exotic species. After the success of cryopreservation technique of somatic cells from endangered animals, cloning is considered as an insurance against further losses in diversity.

3. Research models

Sets of cloned animals could be effectively used to reduce genetic variability and numbers of animals needed for experimental studies. This could be conducted on a larger scale than is currently possible with naturally occurring genetically identical twins. Lambs cloned from sheep selected either for resistance or susceptibility to nematode worms will be useful in studies aimed at discovering novel genes and regulatory pathways in immunology.

4. Human cell-based therapies

SCNT finds direct applications in medicine, principally in therapeutic cloning. Patients with particular diseases or disorders in tissues that neither repair nor replace themselves effectively, like in insulin-dependent diabetes, muscular dystrophy, spinal cord injury, neurological disorders, cancers etc., could potentially generate their own immunologically compatible cells for transplantation which would offer lifelong treatment without tissue rejection.

5. Transgenic animal production

A significant application of cloning is to produce animals from cells that have been genetically modified in order to produce transgenic livestock. Cloning route is regarded as more efficient than conventional pronuclear injection of DNA, where typically less than 1% of injected zygotes develop into transgenic animals. Cloning also enables scientists to introduce, functionally delete or subtly modify genes of interest. Also screening of cells for specific genetic modification before producing transgenic animals is possible in SCNT. It is possible to introduce a specific transgene into desired genetic background of the chosen sex. Cloning enables us to produce embryos or offspring that are all transgenic and none being mosaic. It enables us to produce small herds from each cell line in the first generation, rather than individual founder animals that need to be subsequently bred.

Cloning allow for the careful study of nature-nurture interactions that influence breeding programs by allowing a large enough sample of genetically identical animals to be raised in different environments or with different diets. Such studies have been impossible to perform prior to the advent of SCNT and are likely to yield important information for developing livestock species to live in areas that have, until now, been marginal for food animal production. This is of particular importance to the developing world, where even slightly increased wealth generally favours the incorporation of animal-based agriculture.

Further reading

Betts D, Bordignon V, Hill J, Winger Q, Westhusin M, Smith L and King W. (2001). Reprogramming of telomerase activity and rebuilding of telomere length in cloned cattle. Proceedings of National Academy of Sciences USA, 98:1077–1082.

Bourc'his D, Le Bourhis D, Patin D, Niveleau A, Comizzoli P, Renard JP and Viegas-Pequignot E. (2001). Delayed and incomplete reprogramming of chromosome methylation patterns in bovine cloned embryos. Current Biology,11:1542–1546.

Campbell KH, McWhir J, Ritchie WA.and Wilmut I. (1996). Sheep cloned by nuclear transfer from a cultured cell line. Nature, 380: 64–66.

Chung YG, Mann MR, Bartolomei MS and Latham KE. (2002). Nuclear-cytoplasmic 'tug of war' during cloning: effects of somatic cell nuclei on culture medium preferences of preimplantation cloned mouse embryos. Biology of Reproduction, 66: 1178–1184.

Oback B and Wells D. (2002). Donor cells for nuclear cloning: many are called, but few are chosen. Cloning Stem Cells, 4: 147–168.

Neimann N, Tian X, King W and Lee R. (2008). Epigenetic reprogramming in embryonic and foetal development upon somatic cell nuclear transfer cloning. Reproduction, 135: 151–163.

Chang SC, Tucker T, Thorogood NP and Brown CJ. (2006). Mechanisms of X-chromosome inactivation. Frontiers in Bioscience, 11: 852–866.

Constancia M, Kelsey G and Reik W. (2004). Resourceful imprinting. Nature, 432: 53–57.

Keefer C. (2008). Lessons learned from nuclear transfer (cloning). Theriogenology, 69: 48–54.

Willadsen SM. (1986). Nuclear transplantation in sheep embryos. Nature, 320:63–5.

Kristin W and Prather R. (2010). Somatic cell nuclear transfer efficiency: How it can be improved through nuclear remodeling and reprogramming. Molecular Reproduction and Development, 77:1001–1015.

Ding X, Wang Y, Zhang D, Wang Y, Guo Z and Zhang Y. (2008). Increased pre-implantation development of cloned bovine embryos treated with 5-aza-20-deoxycytidine and trichostatin A. Theriogenology, 70:622–630.

Egli D, Rosains J, Birkhoff G and Eggan K. (2007). Developmental reprogramming after chromosome transfer into mitotic mouse zygotes. Nature, 447:679–685

Okamoto I and Heard E. (2009). Lessons from comparative analysis of X-chromosome inactivation in mammals. Chromosome Research, 17:659–669.

Mayer W, Niveleau A, Walter J, Fundele R and Haaf T. (2000). Embryogenesis Demethylation of the zygotic paternal genome. Nature, 403:501–502.

Tian X, Kubota C, Enright B and Yang X. (2003). Cloning animals by somatic cell nuclear transfer - biological factors. Reproductive Biology and Endocrinology, 1:98.

Tong W, Ng Y and Ng S. (2002). Somatic cell nuclear transfer (cloning): Implications for the medical practitioner. Singapore Medical Journal, 43: 369–376.

Shah R, George A, Singh M, Kumar D, Anand T, CHauhan MS, Manik RS, Palta P and Singla SK. (2009). Pregnancies established from handmade cloned blastocysts reconstructed using skin fibroblasts in buffalo (Bubalus bubalis). Theriogenology, doi:10.1016/j.theriogenology.2008.10.004.

Chapter 12
Embryo Transfer Technology

Introduction

Embryo transfer refers to a step in the process of assisted reproduction, when embryos are placed into the uterus of a female with the intention of establishing pregnancy. The technique, used often in connection with *in vitro* fertilization (IVF), is currently in use in both humans and animals. It is one of the most economical and fastest ways of preserving and multiplying farm animal genetics, especially of cow and buffalo. It is one of the most important reproductive biotechnologies where male and female genetic material could be utilized for the faster improvement of livestock. It has proved to be a powerful technology in genetic improvement of farm animals, primarily to propagate the genes of the females of superior pedigree. With this technology it is possible to exploit the vast reproductive potential of a genetically important cow. While a cow can produce an average of 8 to 10 calves in her entire lifetime under normal management programs, embryo transfer technique can greatly increase the number of offspring. In cattle, particularly in dairy industry, breeding programs have been developed to promote genetic progress by strategic use of elite females through multiple ovulation embryo transfer (MOET) programs.

History of embryo transfer

Most of the applicable embryo transfer technology (ETT) was developed in 1970s and 1980s. ET was first performed and recorded by Walter Heape in 1890. He transferred two Angora rabbit embryos into a gestating Belgian doe, which subsequently gave birth to mixed litter of Belgian and Angora bunnies. ET in food animals began in 1930s with sheep and goats, but it was in 1950s that successful ETs were reported in cattle and pig by Jim Rowson at Cambridge, England. Though the birth of first calf from ET was reported by Willett et al. (1951) at Wisconsin in USA, it was in early 1970's that great commercial interest in bovine embryo transfer developed. European dual-purpose breeds of cattle were imported to North America and because of their relative scarcity became extremely valuable. As a result, there were considerable economic incentives for the application of embryo transfer procedures, as breeders wanted a method to increase the reproductive rate of these females for the profitable sale of their offspring. Thus, the technology of ET in farm animals was developed with money from cattle breeders rather than through traditional research funding. In buffalo, after initial unsuccessful attempts in 1980's, the first successful ET was reported at the University of Florida, USA by Drost et al. in 1983. In India, initial isolated attempts of

ET in dairy animals were made at National Dairy Research Institute (NDRI) Karnal, Central Institute of Research on Buffalo (CIRB), Hisar, Indian Veterinary Research Institute (IVRI), Bareilly, and State Agricultural Universities. At the same time, National Dairy Development Board (NDDB), Anand, initiated a pilot project in 1986 to investigate the utility of the ETT for buffalo improvement. The encouraging results led to launching in 1987 of National Science and Technology Project on "Cattle herd improvement for increased productivity using Embryo Transfer Technology," implemented by the Department of Biotechnology (DBT), Ministry of Science and Technology. NDDB was designated as the lead implementing agency of the project and NDRI, Karnal, National Institute of Immunology (NII), New Delhi, CIRB and IVRI were the collaborating agencies. The main achievements of the project were standardization of superovulatory protocols for cattle and buffaloes with recovery of up to 3.8 viable eggs per flush in cows and 2.8 per flush in buffalo. The conception rate achieved was 38% to 50% following transfer of fresh embryos and 20% and 30% with frozen embryos in cattle and buffalo, respectively. In spite of considerable improvement in ETT, especially during the last two decades, it was felt that poor superovulatory response, improper storage and higher embryo mortality lead to lower conception rate in cloning.

In India, due to ban on cow slaughter almost all the embryos are produced either *in vivo* following superovulation and non-surgical embryo collection or *in vitro* following ultrasound guided trans-vaginal aspiration of oocytes (OPU) and subsequent IVM, IVF and IVC. However, most of the buffalo embryos are produced *in vitro* from the abattoir derived oocytes.

Procedure for embryo transfer

Virtually all commercial embryo transfers performed today uses nonsurgical recovery of the embryos rather than surgical techniques. The process involves several steps and considerable time as well as variable expenses. The various stages of importance are discussed in this section.

1. Selection of donor cow

Reasons for doing embryo transfer on a given animal are more often economic than genetic. Appropriate donor selection will lead to optimal results which would reduce costs and make the procedure much more economical. The donor selection usually involves a previous history of success in embryo transfer. It has been suggested that the potential donor animal be at its prime reproductive age, with a previous history of high level of fertility and superiority in traits of economic importance. The criteria for selecting an outstanding donor cow would thus vary depending on the ultimate purpose of breeding. For example, beef producers take into consideration the performance records, show ring appeal or both, and give significant consideration to potential economic value of the calves. Dairy cattle are selected on one major trait, i.e. milk production. It should however, be noted that ET is not a "cure-all" phenomenon. It does not make average cattle good or good cattle better. It is suitable for a limited number of seed stock producers with beef or dairy cattle that can be bred or species "improvers" for one or more economically important traits. Above all, the potential donor cow should be reproductively sound to produce maximal results. She should have a normal reproductive tract on rectal palpation and have a normal post partum history, especially with regard to cycle lengths of 18 to 24 days. Both beef and dairy cows should be at least 60 days postpartum before the transfer procedure begins. It has been suggested that prospective donor cows in ET programs be selected on the following criteria:

a) Regular heat cycle commencing at a young age.

b) A history of no more than two breedings per conception.

c) Previous calves having been born at approximately 365 day intervals.

d) No parturition difficulties or reproductive irRegularities.

e) No conformational or detectable genetic defects.

The donor cow should be maintained at the level of nutrition appropriate for her size and level of milk production. Both very obese and thin cow will have reduced fertility. Thus, it is important that the donor cow be in an appropriate body condition score at the time of embryo transfer. The strict selection criteria will not only ensure genetic superiority but should also ensure a high level of success thereby making the procedure more economical.

2. Superovulation of the donor cow

Superovulation of the donor cow is the next step in ET process. It refers to release of multiple eggs at a single estrus. Superovulation of the donor animal is the most critical step of the technique of *in vivo* embryo production, as it directly affects the yield of embryos per donor. The basic principle of superovulation is to stimulate extensive follicular development through the use of a hormone preparation with follicle stimulating hormone activity, which is given intramuscularly or subcutaneously. Usually the commercially available preparations of FSH are injected twice daily for four days at the middle or near end of normal estrous cycle, when a functional corpus luteum (CL) is on the ovary. A prostaglandin injection is given on the third day of the treatment schedule which will cause CL regression and a heat or estrus to occur approximately 48 to 60 hours later. A number of comparative studies have been performed to ascertain the efficacy of FSH or equine chorionic gonadotrophin (eCG), their dosage and route of administration, effect of pre-treatment of donor animals as well as the ideal day (s) of estrous cycle for superovulation. The comparative studies of superovulation with FSH/eCG suggested that FSH (follitropin) induces better superovulatory response recovery of significantly more transferrable embryos (4.4–4.7 vs 0.7–3.2) and viable embryos (2.3–3.1 vs 0.6–2.2). It has also been found that 5 day treatment regimens of FSH give better superovulatory response and embryo recovery as compared to FSH administered twice daily for 4 days. In cattle and buffalo, since the advent of PGF2α, gonadotropins are generally administered during mid cycle when the CL is fully functional and progesterone (P4) concentration is highest, as its level at the time of initiation of superovulation treatment has positive correlation with superovulatory response in buffalo. However, the short life of FSH necessitates its intramuscular administration twice daily for 3 to 5 days, resulting in repeated handling of donor which is not only stressful but requires repeated travelling to considerable distances if the donors are located at different places in the rural areas. To overcome this problem, attempts were made to regulate slow absorption of FSH by injecting total dose of FSH as a single dose subcutaneously. But it was found that subcutaneous administration of FSH resulted in poorer superovulatory response and embryo recovery than intramuscular route.

3) Insemination of the cow

Since many ova are released from the multiple follicles of the ovary following superovulation, there is greater than normal need to be certain that viable sperm reach the oviducts of the superovulated females. Therefore, many embryo transfer technicians choose to inseminate the cow several times during and after estrus. One scheme that has been used successfully is to inseminate the superovulated cow at 12, 24 and 36 hours after the onset of standing heat. Thus, use of high quality semen with high percentage of normal, motile cells is a very critical step in any embryo transfer program. The correct site for semen placement is in the body of the uterus. This is a small target (1/2 to 1 inch) that is just in front of the cervix. There seems to be a tendency for the inseminators to pass the rod too deep and deposit the semen into one of the uterine horns, thereby reducing fertility if ovulations are taking place at the opposite ovary.

4. Embryo flushing

Knowledge of the time of arrival of embryos in the uterus and their subsequent development is necessary to decide the time of collection of embryos of desired stages. It has been demonstrated in buffalo that nearly all the embryos reach uterus around day 5.25 after the onset of superovulatory estrus and most of the early stage (4 cell) embryos develop to 8 to 16-cell/ early morula by day 5, morula/compact morula by day 5.5 and to blastocyst by day 6. Thus, the ideal time for nonsurgical embryo collection is day 5.5 to 6 after estrus. Buffalo seems to differ from superovulated cattle, as in the latter species the ideal time for nonsurgical embryo collection is day 7±1 after superovulatory estrus. Embryos could be collected either surgically or non-surgically but most of the ET employs non-surgical procedure. To collect the embryos non-surgically, a small synthetic rubber catether is inserted through the cervix of the donor cow, and embryo flushing medium is flushed into and out of uterus to harvest the embryos. This collection procedure is relatively simple and can be accomplished in 30 min or less without harm to the cow. In this procedure, a pre-sterilized stylet is placed in the lumen of the catheter to offer rigidity for passage through the cervix into the body of the uterus. When the tip of the catheter is in the body of the uterus, the cuff is slowly filled with approximately 2 ml of normal saline. The catheter is then gently pulled so that the cuff is seated into the internal os of the cervix. Additional saline is then added to the cuff to completely seal the internal os of the cervix. A Y-connector with the inflow and outflow tubes is attached to the catheter. A pair of forceps is attached to each tube to regulate the flow of flushing fluid. The fluid is subsequently added and removed by gravity. The fluid in the uterus is agitated rectally, especially in the upper one-third of the uterine horn. The uterus is finally filled with medium to about the size of a 40 day pregnancy. About 1L of fluid is used per donor. Many operators use a smaller volume and flush one uterine horn at a time. Each uterine horn is filled and emptied five to ten times with 30 to 200 ml of fluid each time, according to size of the uterus. The embryos are flushed out with this fluid into a large graduated cylinder. After about 30 min, embryos settle and can be located under a stereomicroscope by searching through an aliquot from the bottom of the cylinder.

5) Evaluation of the embryos

The individual embryos are located using a microscope and evaluated for their quality for numerical classification of their potential likelihood of success if transferred to a recipient female. The major criteria for evaluation include:

i) Regularity of shape of the embryo

ii) Compactness of the blastomeres.

iii) Variation in cell size.

iv) Colour and texture of the cytoplasm

v) Overall diameter of the embryo

vi) Presence of extruded cells

vii) Regularity of zona pellucida

viii) Presence of vesicles (small bubble like structures in the cytoplasm).

Embryos are classified according to these subjective criteria as follows:

Grade 1: Excellent or Good

Grade 2: Fair

Grade 3: poor

Grade 4: Dead or degenerating.

The embryos are also evaluated for their stage or development without regard to quality. These stages are also numbered:

Stage 1: Unfertilized

Stage 2: 2 to 12 cell

Stage 3: Early morula

Stage 4: Morula

Stage 5: Early blastocyst

Stage 6: Blastocyst:

Stage 7: Expanded blastocyst

Stage 8: Hatched blastocyst

Stage 9: Expanding Hatched Blastocyst

There is apparently no difference in pregnancy rates of fertilized cells in different stages of development assuming that they are transferred to the recipient female at the appropriate stage of estrous cycle. Stage 4, 5 and 6 embryos endure the freezing and thawing procedures with greatest viability. Embryo quality is also of utmost importance in the survival of the freezing and thawing stress. Grade 1 embryos are generally considered only ones to freeze. Grade 2 embryos can be frozen and thawed, yet pregnancy rates typically are reduced.

6) Selection and preparation of recipient females

The selection of proper recipients and their management is critical to embryo transfer process. Cows that are reproductively sound, exhibit calving ease and have good milking and mothering ability are prospective recipients. They must be on a proper plane of nutrition (body condition score 6 for beef cows and 3 to 4 for dairy breed recipients). They must also be on sound herd health program. To maximize embryo survival in the recipient female following transfer, conditions in the recipient reproductive tract should closely resemble those in the donor. This requires synchronization of the estrus cycles between the donor and the recipients, optimally within one day of each other. Synchronization of the recipients can be done in a similar manner and at the same working time as the donor cows. Prostaglandins should be injected into recipients at the same time they are injected into donor cow, as it optimizes the probability that the recipient will be in the same stage of the estrous cycle as the donor cow. "Synchro-Mate-B" system, involving injection of recipients and implantation with a synthetic progesterone, has been used successfully. The implant is removed nine days after its insertion and the cows will show standing estrus approximately 30 to 40 hours later. This timing again must match the time of insemination of the donor cow so that the donor and the recipients have a similar uterine environment seven days later when the transfer takes place. Synchronizing drugs are only effective on recipient females that are already cycling. Anestrus or non-cycling cows that are too thin or too short in postpartum days will not make useful recipients.

7) Transfer of the embryos

Untill very recently, most embryo transfers in the cow were done surgically, whereas at present most are done by nonsurgical methods. Surgical transfers were done initially by way of mid-line incision which necessitated a general anesthetic and rather elaborate facilities. During the mid to late 1970's, surgical transfers were done by way of flank incision which was quicker and did not necessitate the same sophistication in facilities. This made 'on farm' embryo transfer possible and added a whole new perspective to use of embryo transfer production schemes. More recently, the use of nonsurgical embryo transfer has increased the utilization of this technology in cattle breeding schemes because of even less elaborate requirements. In surgical transfer, the uterine horn adjacent to the ovary bearing the corpus luteum (CL) is exteriorized and embryo is deposited through the uterine wall into the uterine lumen with a Pasteur pipette or an intravenous catheter. Nonsurgical embryo transfer techniques involve the use of Cassou AI pipette or some similar apparatus. The embryo is placed in the uterine horn adjacent to the ovary bearing the CL by passing the pipette through the cervix, very similar to AI. It has been observed that practice and dexterity improve the ability to achieve high pregnancy rates, suggesting that trauma to the endometrium might be a limiting factor in this method of embryo transfer, while stimulation of the cervix and inadvertent introduction of bacterial contaminants do not seem to be major determinants under normal circumstances. Embryo flushing and embryo transfer are both done after an epidural anesthetic to block contractions of the digestive tract and aid in manipulation of cervix and uterine horns.

Cost of embryo transfer

The costs of embryo transfer are highly variable, as many different options and packages are offered by embryo transfer technicians. Some technicians perform embryo transfer only on the farm or ranch where the donor cow is located, while others have facilities to house and board donor and recipient cows and perform embryo transfer under hospital-like conditions. The minimum cost of $250 has been reported by most of the embryo transfer technicians. The cost may not include drug costs for superovulation, and for semen, registration, embryo transfer certificates, blood typing of donor cows and ancestors and the cost of maintaining the recipient cow until the calf is weaned. Three to five straws of valuable semen can be priced from $45 to $300. The proper health care, nutrition and synchronization of donor and recipient can add another $400 to $500 expense to each successful pregnancy. Thus, many purebred operations conducting embryo transfer on a Regular basis consider that each ET calf must have a market value of $1500 to $2000 greater than other naturally conceived and reared calves in the herd before embryo transfer is considered.

Factors affecting superovulatory response

A range of factors, as diverse as animal factors, environmental factors as well as technical factors affect superovulatory response. Among them the most important are season, breed of animal, dominant follicle, repeated superovulation, various pre-treatments and bulls used for AI. This section gives a brief description of all these factors:

i) Season

Seasonal variation in reproductive functions in cattle and buffalo is greatly related to heat stress during summer as they are unable to dissipate environmental heat. This in turn increases the temperature of microenvironment of embryo (oviduct and uterus) which affects the embryonic viability adversely. A poor superovulatory response to gonadotrophin in summer has been observed in buffaloes, than other seasons. It has also been observed that the embryos received following superovulation in dry hot seasons were not of transferable grade. In cattle also more mean ovulations have been reported in rainy season than winter and summer.

ii) Age/ parity

Decreased fecundity with advanced age has been reported in cattle and other mammalian species. Fertilization rate as well as embryo quality has also been found to get deteriorated with age. However, in buffalo it has been observed that mean ovulation rate in the nulliparous buffaloes up to third parity and more than third parity was approximately, 6.59, 6.82 and 5.98, respectively, suggesting no significant effect of parity on superovulatory response.

iii) Breed

Superovulatory response and embryo recovery varies between breeds of cattle, different laboratories and individuals. Zebu cattle were found to be the lowest producers of eggs following superovulatory response but had the best pregnancy rates (56%) after ET.

iv) Effect of dominant follicle

Presence of dominant follicles at the time of superovulation influences superovulatory response. In cows, superovulatory response in absence of dominant follicle was better than in its presence. It has also been reported that manipulation of dominant follicle by hCG, estradiol valerate or GnRH failed to improve the superovulatory response.

v) Repeated superovulation

It is desired that the donor must respond optimally to repeated superovulation, especially when large number of calves are to be produced in open nucleus breeding system. A drop in embryo production has been observed in buffaloes after first and second superovulation. However, the repetitive superovulatory treatments did not affect the fertility of the donors as 83.3% buffaloes conceived with only 1.8 services per conception.

vi) Effect of various pretreatments

Various pretreatments (hCG, estradiol, eCG, GnRH) have been attempted to enhance the superovulatory response in cattle and buffaloes. However, pretreatment with 250 IU hCG (i/m) on ninth day of estrous cycle and 3 mg estradiol valerate (sc) on tenth day of estrous, before superovulation with FSH failed to improve the superovulatory response and embryo recovery in cattle. An attempt to increase more follicle recruitment in donor buffaloes by pre-treatment with 4–5 mg FSH on day 3 and 4 of the estrous cycle before superovulation on day 10 and 11, has also been reported as failed in context to ovulation rate and embryo recovery.

vii) Effect of bull on the quality of embryos

Semen quality and dose does play a significant effect on embryo recovery as well as quality. A study comparing semen from 5 bulls showed that percent viable embryo recovery varied considerably, ranging from 32.6 to 84.4%. The mean number of unfertilized ovum and percent unfertilized ovum recovered varied significantly between bulls.

Applications of embryo transfer technology

1. Genetic improvement

Genetic progress is considered to be slower through the use of embryo transfer than it is using conventional artificial insemination, especially on a national herd basis. However, with increased selection intensity and shortened generation intervals, i.e., by transferring female offspring, genetic gain can be made on a within herd basis. The production of about six offspring per donor cow could double selection intensity and the rate of response to genetic selection for traits such as growth that can be measured in both sexes. This could especially be worthwhile in improving elite herds, the genetics of which could be spread over a large population through use of AI. Although the rate of genetic improvement in dairy cattle may range from 2.5%–8%, it could be increased 3 to 4 times through ETT, especially in single offspring bearing animals like cow and buffalo, if dairy replacements were selected from the top 10% of the herd. Thus, the application of ETT is highly desirable to produce AI bulls from the best proven bulls and cows or buffaloes available.

2. Planned matings

By far the most common use of embryo transfer in animal production programme is the proliferation of so-called desirable genotypes. AI has permitted the widespread dissemination of a male's genetic potential, while ET provides the opportunity of disseminating the genetics of proven elite females. Embryo transfer also permits the development of herds of genetically valuable females, most of which may be sibs if not full-sibs. As AI has led to the very valuable bull, now embryo transfer has resulted in the very valuable female. Many breeders have identified individual females whose offspring are most saleable and used them exclusively for embryo transfer. Embryo transfer has been used to rapidly expand a limited gene pool.

3. Genetic testing for Mandelian recessive traits

A common use of embryo transfer procedures is to genetically test AI sires for deleterious heredity traits. Some AI organizations keep carriers of certain genetic defects on one hand to serve as donors in testing new sires. Embryos are transferred into unrelated recipients and pregnancy may be terminated at various stages to examine fetuses for presence or absence of the defect. Depending on the heritability of the defect, generally eight to ten fetuses are sufficient to declare a bull free of trait. Another alternative is to mate the bull in question to seven or eight of his superovulated daughters. The offspring will then represent all recessive traits that a bull may carry. A less attractive alternative would be to naturally mate the bull in question to 40–50 of his daughters.

4. Twinning in cattle

Beef production is currently regarded as less efficient as not all the cows produce a calf each year. However, it has been estimated that unit beef production can be increased by 60% in intensively managed herds through twinning. It has been estimated that 70% of nutrient intake by a beef cow is utilized for her own maintenance, whereas only 30% is for growth and maintenance of calf during pregnancy and lactation. Thus, it feels desirable to take advantage of the efficiency of gestation and lactation. Genetic selection for twinning in cattle has been largely unsuccessful and gonadotrophin treatment to induce twinning has also been unreliable. Embryo transfer does provide a very real alternative in the production of twins. The most limiting factor at this time is the cost of transfer which continues to exceed the average price for calves to be raised for meat. The nonsurgical transfer of a previously frozen embryo to a recipient that has already been serviced may become an economical method of producing twins. Other alternatives include the nonsurgical transfer of two or three frozen embryos to commercial beef cows. However, the recipients carrying twins require extra nutrition and management, especially around calving. The recipients must be sufficiently large to carry twins and produce enough milk for their feeding. The freemartins also make twinning unattractive for pure breeding purposes. The incidence of freemartins is luckily relatively low and sexing of embryos would eliminate the problem entirely. Similarly, the production of identical twins by microsurgery would make twinning very feasible for embryo transfer programmes in purebred herds.

5. Disease control

It is an established fact that infectious diseases in bovines and buffalo are unlikely to be transmitted by the embryo. Thus, it has been suggested that embryo transfer be used to salvage genetics in the face of disease outbreak. However, much research has to be done on embryo-virus interactions before this technique can be carried out with complete confidence.

6. Import and export

The intercontinental transport of a live animal may cost $1000 or more, whereas an entire herd can be transported, in the form of frozen embryos, for less than the price of a single plane fare. This is regarded as the single most potential application of embryo transfer. Additional benefits include a wider genetic base from which to select and the retention of genetics within the exporting country and adaptation. This is particularly true for tropical and subtropical climates where the embryo would have an opportunity to adapt both in the uterus and then suckling a recipient indigenous to the area. However, well defined methods of collection, handling and washing embryos must be followed to ensure that disease transmission in avoided.

7. Salvage of reproductive function

Embryo transfer procedures have been useful in the diagnosis, treatment and salvage of reproductive function in so called infertile cows. Although it is recommended that the cause of infertility must not be of genetic origin, but this is often difficult to determine. Another important use of embryo transfer is to salvage the genetics of terminally ill animals. It may be possible to produce an additional two or three offspring through embryo transfer before animal dies.

8. Research

Embryo transfer techniques have proven to be a very useful research tool. In fact, many technical developments in embryo transfer prior to 1970 were directed towards research purposes rather than for the propagation of superior livestock. These studies included natural limitations to twin pregnancies, uterine capacity, endocrine control of uterine environment, maternal recognition of pregnancy, embryo-endometrium interactions and endocrinology of pregnancy. The production of identical twins, clones, chimeras etc. will certainly advance many of these sciences.

Embryo cryopreservation

One of the pillars for success of embryo transfer is efficient and successful cryopreservation. Successful embryo freezing leads to better recipient management and makes ETT more cost-effective. In addition, season of parturition can be controlled, even though embryo collection and freezing may take place around the year. ET also allows rapid and efficient progeny and performance tests of sibs. Further, full sibs or identical sibs can be frozen until genetic worth of those transferred can be established. Embryo freezing is necessary for international movement of embryos because it eliminates critical timing and allows disease testing while the embryos are held in quarantine.

Cellular freezing constitutes a complex physiochemical process of heat and water transport between the cell and its surrounding medium. One of the important factors affecting embryo freezing is the cooling rate. It is dependent on the size of the cell, its surface to volume ratio, its permeability to water and the temperature coefficient of that permeability. Cells are injured during freezing and thawing primarily by solution effects and intracellular ice formation. At high cooling rates the dehydration of the cell falls behind that of the solution, so an intracellular ice forms. To avoid intracellular freezing, embryos must be cooled at 1°C/min or slower. The too low a rate of cooling can also damage cells by what has been referred to as solution effect. This is especially harmful if cells are not allowed to rehydrate during thawing. Thawing rate is also one of the detrimental factors for embryo survival and depends on the freezing regimen used. Two methods are generally used for cryopreservation: *slow freezing-rapid thawing* and *vitrification*. Slow freezing protocols are straightforward and can be easily applied for cryopreservation of embryos. In addition, more number of embryos could be frozen in one vial. It has been observed that the survival of embryos subjected to slow freezing–rapid thawing is poor and quite often requires a programmable freezing module. Vitrification, on the other hand, is less expensive as it does not involve expensive programmers. This technique is also far more time-efficient, requiring only several minutes as compared to 1–2 hours required for controlled slow freezing. The embryos are normally stored in liquid nitrogen at -196°C. The only reaction that occurs at this temperature is direct ionization from the background radiation. Thus, storage times of 200 years or so are likely to produce any detectable reduction in the survival of frozen embryos or cause genetic change. Cryoprotectants such as Glycerol or DMSO in concentrations ranging from 1 to 2 M are required to ensure embryo survival after freezing. It is presumed that cryoprotectants act by reducing the amount of ice present at any temperature during freezing, thereby moderating the changes in solute concentration. During the addition and dilution of a permeating cryoprotectant, the cell undergoes osmotic changes in cell size, thus it is to be ensured that the addition or dilution is not carried out inappropriately. The rate of removal of the cryoprotectant is also critical. The standard empirical method is to dilute it by step-wise addition of PBS or to pipette the embryos into decreasing concentrations of cryoprotectant solution, e.g., 0.25 M steps. Another modification has been inclusion of nonpermeable solutes like sucrose into the dilution medium which acts as an osmotic counterforce to restrict water movement across the membranes. As the cryoprotectant leaves the embryo, it will shrink in response to the extracellular hypertonic dilution medium. It regains its normal volume when at the end of the process the embryo is placed in normal isotonic culture medium.

Further reading

Adams CE. (1982). Egg transfer: historical aspects. In: Adams, C.E. (Ed.), Mammalian Egg Transfer. CRC Press, Boca Raton, FL, pp. 1–17.

Anderson GB. (1978). Methods of producing twins in cattle. Theriogenology, 9: 3 -16.

Betteridge K. (2003). A history of farm animal embryo transfer and some associated techniques. Animal Reproduction Science 79: 203–244.

Betteridge KJ. (1977). Embryo Transfer in Farm Animals: A Review of Techniques and Applications. Monograph No. 16. Canada Department of Agriculture, Ottawa.

Bradford GE and Kennedy BW. (1980). Genetic aspects of embryo transfer. Theriogenology, 13: 13–26.

Hasler F. (2003). The current status and future of commercial embryo transfer in cattle. Animal Reproduction Science 70: 245–264.

Land BR and Hill WG. (1975). The possible use of superovulation and embryo transfer in cattle to increase response to selection. Animal Production, 21: 1–12.

McDaniel BT and Cassell BG. (1981). Effects of embryo transfer on genetic change in dairy cattle. Journal of Dairy Sciences, 64: 2484–2492.

Misra AK, Shiv P and Taneja VK. (2005). Embryo transfer technology (ETT) in cattle and buffalo in India: A review. Indian Journal of Animal Sciences, 75: 842–857.

Sreenan JM. (1983). Embryo transfer procedure and its use as a research technique. Veterinary Record, 112: 494–500.

Widayati DT. (2012). Embryo transfer as an assisted reproductive technology in farm animals. World Academy of Science, Engineering and Technology, 6:10–21.

Chapter 13

Stem Cell Technology in Farm Animals

Introduction

Almost three decades from now, a breakthrough was achieved when murine embryonic stem cells were isolated from the inner cell mass of developing blastocysts and grown *in vitro* under usual laboratory conditions. Their two properties: i) capability to be propagated indefinitely in culture and, ii) ability to contribute to all cell lineages, including germ line, when incorporated into chimeras with intact mouse embryos; left the then scientific world bewildered and in excitement. A cell is regarded as a stem cell if it fulfills the essential characteristics of self-renewal, capability for multilineage differentiation and *in vivo* functional reconstitution of a given tissue. Cells derived from many different sources have been shown to fulfill these criteria and hence are considered as stem cells. Based on their source of origin, these could either be adult stem cells, embryonic stem cells, embryonic germ cells or placental stem cells. These cells have different potencies and are accordingly classified as totipotent, pluripotent, multipotent or unipotent. The embryonic stem cells are derived from the inner cell mass of a blastocyst, and adult stem cells are isolated from the tissues or organs of the body.

Farm animal stem cell research

It was presumed that the embryonic stem cells would provide a simple model system to study the basic processes of early embryonic development and cellular differentiation, and promised for cell-based therapies for humans if and only if human embryonic stem cell lines could be developed. This would provide a radical new approach to treatment of a variety of disease in which organ damage exceeds the body's natural repair capability. A genuine belief that stem cell research will deliver a revolution in terms of treatment for cardiovascular disease, neurodegenerative disease, cancer, diabetes, and the like, propelled the research at such a pace that the feat was accomplished within a decade with the publication of two papers describing the growth *in vitro* of human embryonic stem cells derived either from inner cell mass (ICM) of early blastocyst or primitive gonadal regions of early aborted foetuses. In the mean time, extensive studies carried out world-wide in stem cell research demonstrated the usefulness of these cells for introduction of both dominant and recessive mutations into murine germ line. These studies together with the ability for production of unlimited number of cells, while still retaining developmental potential, provided a strong incentive for isolation of domestic animal embryonic stem (ES) cells. The speculations that development of ES cell lines of domestic livestock, such as cattle, buffalo, sheep and goat, would greatly facilitate precise genetic manipulation for better health, increased disease resistance, increased milk production and desired composition, increased growth rate with improved carcass composition, enhanced reproductive performance and prolificacy, demonstrated their usefulness at par with those of human ES cell lines. The potential to produce transgenic animals using ES cells and their use as models for human diseases, cell transplantation therapies and studies of lineage commitment and development further added to the zeal. In addition, the exploitation of these cells for biopharming, production of recombinant proteins for treatment and prevention of human and animal diseases and development of genetically modified animals to serve as bioreactors with capability of secreting commercially important mammalian proteins in their fluids added to their further usefulness and thereby scientific research. But before these potentials could be realized a large number of problems had to be overcome. The first and foremost was to develop the means of producing a sufficient number of healthy embryos under laboratory conditions followed by development of efficient methodologies to isolate ES cells from small number of such elite embryos. The second difficulty was to obtain and maintain the ES cell lines which would reproductively contribute to germ line. This was further compounded by differences between mouse and farm animals in reproductive behaviour and generation time. Because of shorter generation time in mice, it is possible to produce chimeras that can be bred to obtain mice which are heterozygous and then homozygous for the introduced mutation. Since, domestic animals have a longer generation time and it is desirable to maintain the elite genetic background of such ES cells, techniques which will produce 100% ES cell-derived animals had to be developed. Despite all these challenges, ES cell research in domestic animals proceeded at a pace, next only to human ES cell research and within a span of time we had ES cell lines or ES cell- like lines for bovine, ovine, caprine, equine and porcine species. Our lab was the first to develop ES cell - like cells from *in vitro* produced buffalo embryos that survived up to eight passages only. From there on, continuous research in our lab resulted to development of fully characterized ES cell lines of bubaline species, derived from the inner cell masses of *in vitro* fertilized, parthenogenetic, as well as Hand-guided cloned (SCNT) embryos. We have been successful in maintaining ES cell lines derived from IVF- embryos beyond 100 passages. Furthermore, we have been able to optimize their culture conditions, decipher their signaling pathways, analyze the expression pattern of pluripotency markers, and study their differentiation strategies into desired cell types. Our current research focuses on use of these cells for transgenic animal production either through our in-house developed Hand- guided cloning (HGC) technique or through their directed differentiation into germ line cells. Using HGC technique, we have already been able to produce world's first buffalo calf (Garima II) using ES cell as the donor cell. The calf is in a sound physiological and reproductive health and has recently given birth to a normal calf (Mahima) through artificial insemination. We have also been successful in differentiation of these stem cells into germ cells, sperm and oocytes. We were further able to produce haploid cell population in our differentiation strategies and have developed a differentiation cocktail (Syed-Chauhan Medium, SCM) for the purpose. We successfully engineered VASA-transgenic buffalo embryonic stem cell lines for differentiation to germ cells.

One of the challenging problems facing stem cell research is the accumulation of mutations while the cells are in culture and the instability of XX murine ES cell lines, and the cell lines of other species also, as seen by some researchers. Thus, it becomes essential to perform routine karyotyping, if necessary the cytogenetic studies also, of ES cell lines to check for chromosomal aberrations.

The belief that early (human) embryos holds the same status as any sentient being puts 'harvesting' of stem cells from blastocysts in morally unjustifiable activity. This presses for other sources of malleable stem cells to be sought. This prompted the search for adult stem cells and induced pluripotent stem cells (iPSc). The relatively difficult procedures of *in vitro* embryo production further make establishment of embryonic stem cells a difficult task, thereby indirectly favouring adult stem cell and iPSc derivation. In an adult, organ formation and regeneration was thought to occur through the action of organ- or tissue-restricted stem cells (i.e. haematopoietic stem cells giving rise to all the cells of the blood, neural stem cells making neurons, astrocytes, and oligodendrocytes). However, it is now believed that stem cells from one organ system, for example the haematopoietic compartment can develop into the differentiated cells within another organ system, such as the liver, brain or kidney. Thus, certain adult stem cells may turn out be as malleable as ES cells and so also be useful in regenerative medicine. With this developed a whole new classification of embryonic stem cells, adult stem cells, embryonic germinal cells, fetal stem cells, placental stem cells, umbilical cord stem cells, etc. These cells vary in potency from pluripotent to unipotent through various hierarchies of multipotency. In this section, we will be discussing the derivation of buffalo ES cell lines, their maintenance and propagation in culture, their characterization, applications, challenges and future prospects.

Derivation of buffalo embryonic stem cells

Initial research into the isolation of domestic animal ES cells in our and other laboratories attempted to repeat the work carried out in mice by isolating cell lines directly from cultured preimplantation embryos. ICM is removed from the blastocysts, either mechanically, enzymatically or by applying antibodies (immunosurgery), and cultured on appropriate feeders to obtain a stem cell line. The feeders most often consist of homologous feeder cells treated with Mitomycin C. In our laboratory, we isolate ICMs of hatched blastocysts, derived either from IVF, Hand-guided cloning or parthenogenesis, by mechanical cutting with the help of Microblades™ under a zoom stereomicroscope (Olympus, SZ40, Japan). The isolated ICMs are then seeded individually and separately on the feeder layers (buffalo fetal fibroblasts treated with 10µg/ ml Mitomycin C for 3 h (Fig. 13.1), and cultured in embryonic stem cell media.

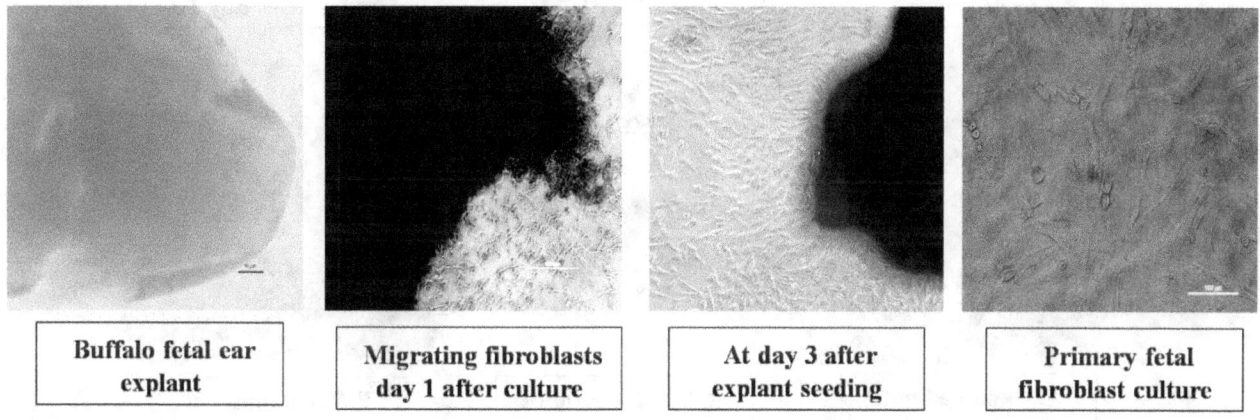

| Buffalo fetal ear explant | Migrating fibroblasts day 1 after culture | At day 3 after explant seeding | Primary fetal fibroblast culture |

Fig. 13.1: Establishment of buffalo fetal fibroblast culture

The media which has been optimized for maximum stem cell growth and maintenance of pluripotency is composed of KODMEM and 15% Knockout serum replacer (KSR) supplemented with 2mM L-Glutamine, 5ng/mL basic fibroblast growth factor 2 (bFGF-2), 1000U/ml recombinant murine leukemia inhibitory factor (rmLIF), 1X nonessential amino acids and 50µg/ml gentamicin sulphate and incubated at 37–38°C in 5% CO_2 incubator. The culture medium is changed on alternate days and further colonization of the cells is observed routinely under an inverted microscope (Fig. 13.2). The primary colonies obtained 8–10 days after seeding of ICMs are disaggregated with the aid of Microblade under a zoom stereomicroscope. Aggregates of cells are individually reseeded onto new feeder layers. The colonies exhibiting typical morphological features of ES cell-like cells are subcultured using mechanical dissociation until the cells remained in an undifferentiated state or when colony formation stops. The dissociated colonies are subcultured on to new feeder layers and further passaged after every eight days onto the new feeders and so on. Some of the colonies are randomly chosen for characterization for embryonic stem cell markers, karyotyping and embryoid body formation. After confirming their stemness, the colonies are cryopreserved at different passages so as to maintain the stock, while the remaining colonies are propagated for further studies.

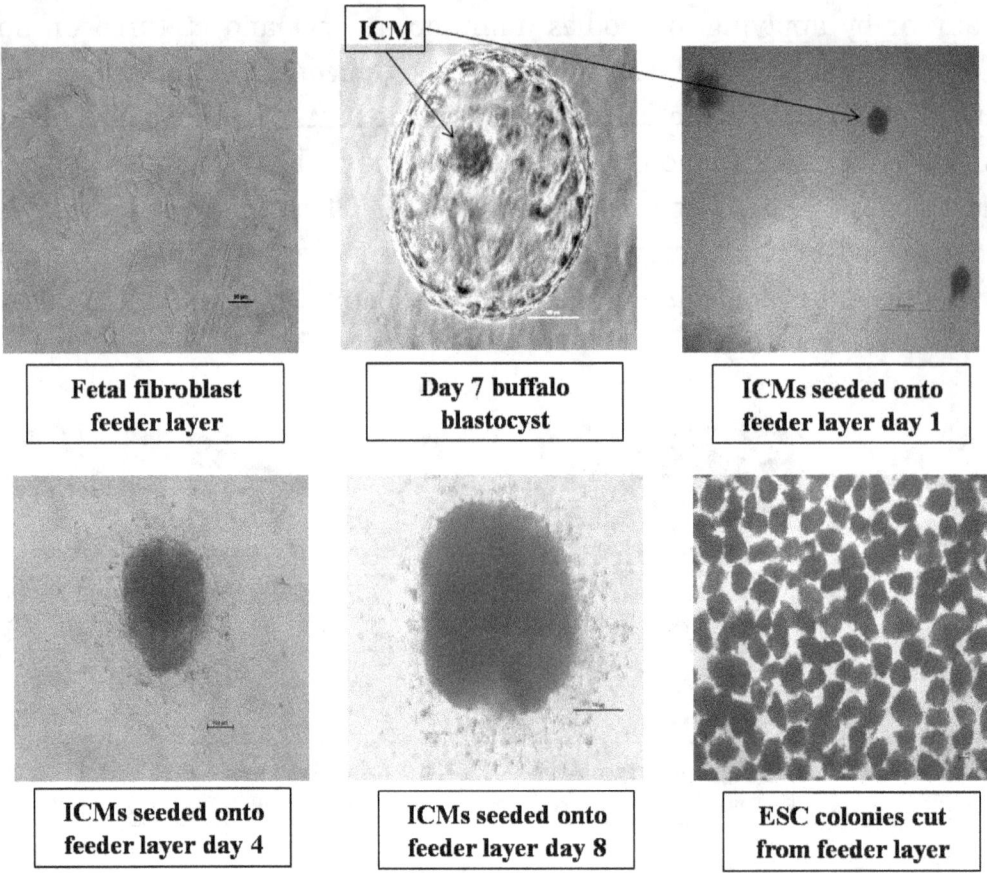

Fig. 13.2: Establishment of buffalo embryonic stem cell culture

ES cells, in general, can also be grown under feeder-free conditions using dishes coated with animal-based ingredients with the addition of MEF-cell (Mouse embryonic feeder) conditioned medium. Such systems are not optimal for derivation and growth of clinical grade human ES cell lines since they bear a risk of cross transfer of infectious agents. There are also reports of high differentiation rates of these cell lines and genomic instability after prolonged *in vitro* growth, revealing the limitations of applied feeders or feeder free systems for growth of undifferentiated human ES cell lines. When plated on feeder layers in appropriate culture medium, the ICM clumps are allowed to adapt to the new culture conditions and get attached to the feeder cells and start growing. It has been observed that ICM clumps with 50–100 cells have more chances to attach and proliferate than smaller cell clumps and for successful derivation of the cell lines the ICMs should be plated as clumps and not as single cells, since ICM cells are held together tightly with numerous junctional complexes.

Characterization of embryonic stem cells

Given the historical introduction of the term "ES cell" and the properties of mouse ES cells, the essential characteristics of primate ES cells should include (i) derivation from the pre-implantation or peri-implantation embryo, (ii) prolonged undifferentiated proliferation, and (iii) stable developmental potential to form derivatives of all three embryonic germ layers even after prolonged culture. For ethical and practical reasons, in many primate species including humans, the ability of ES cells to contribute to the germ line in chimeras is not a testable property and hence, is not considered as a pre-requisite for ES cell characterization. The mouse ES cells provides a benchmark for definition of the generic features that any ES cell might be expected to possess and other properties which may be peculiar to bonafide pluripotent cells isolated from different species or different tissues, or representative of a different stage of embryonic development. Its key features include: derivation from a pluripotent cell population; stably diploid and normal karyotype *in vitro*; ability to be propagated indefinitely in the primitive embryonic state; capability of spontaneous differentiation into multiple cell types representative of all three embryonic germ layers, both in teratomas after grafting or *in vitro* under appropriate conditions; potential to give rise to any cell type in the body, including germ cells, when allowed to colonize a host blastocyst. The criteria for pluripotency usually include derivation of the stem cell line from a single cloned cell which eliminates the possibility that several distinct committed multipotential cell types are present in the culture that together account for the variety of differentiated derivatives produced. To ensure that the line so established is a true stem cell line various attributes like morphological resemblance, pluripotency related marker expression, growth requirements and self renewal capability are thus taken into account. To characterize our buffalo ES cell lines, as true ES cell lines, we adhered to following established criteria in our laboratory:

a) Morphology

Mouse and human ES cell colonies have a characteristic morphology. They usually proliferate in tight round shaped colonies with smooth edges. The morphology of ES cells has two important traits–they have quite small amount of cytoplasm and they exhibit faster proliferation rate in a given population of cells. However, for many animals such as cow, sheep, pig, horse, hamster, mink, rabbit, primates, embryo derived ES cells are found to propagate as flattened colonies, almost as a monolayer with individually distinct cells that have been described as epithelial-like or epitheloid, whereas in case of buffalo the primary colonies have been observed to be dome shaped with abundant lipid-like vacuoles (Fig. 13.3).

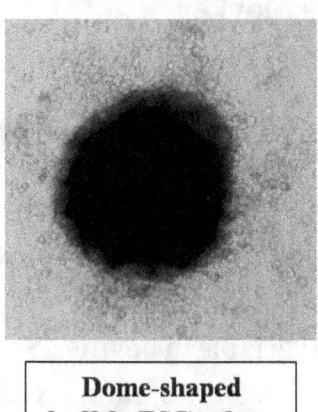

Dome-shaped buffalo ESC colony

Fig. 13.3: Morphology of buffalo embryonic stem cell colony

b) Cytogenetic Analysis

Karyotyping of ES cells has recently received much attention in most laboratories. This has been studied by G-banding with most ES cells exhibiting a normal compliment of chromosomes. Karyotyping is generally performed at different passages in order to know the genetic stability of ES cells during culture. We also performed karyotyping at different passages in randomly collected colonies and found that the cells retained the normal chromosomal composition (48 +XX in our case) throughout the culture interval (Fig. 13.4).

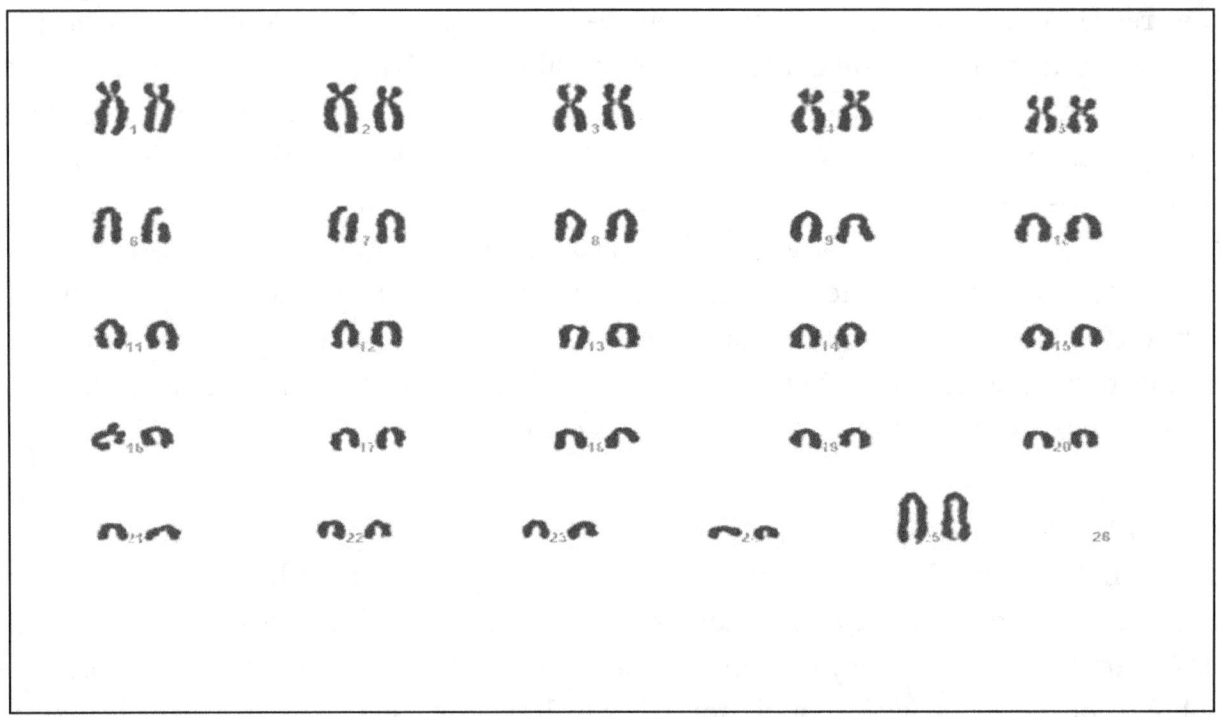

48+XX Karyotype of buffalo ES cells

Fig. 13.4: Karyotype of buffalo embryonic stem cell line

c) Expression of Various Surface Markers

Alkaline Phosphatase

Alkaline phosphatase (ALP) enzyme is secreted by almost all the cells but its intensity is higher in undifferentiated cells. ALPs located at the cell surface are linked to the cell membrane *via* a phosphatidylinositol glycan linkage. ES cells are known to express a high specific activity of ALP, which declines during progressive differentiation resulting in low ALP activities in differentiated cells. Mouse ES cells display ALP activity but the signal for this pluripotent cell marker seems to be variable in bovine ES cell-like cells. ALP has been used as a marker for porcine ES cells, sheep ES cell lines, canine stem cell-like cells and buffalo ES cell-like cells and by many groups in bovine as an indicator for ES cells. Our buffalo embryonic stem cell colonies also exhibited a high alkaline phosphate activity (Fig. 13.5).

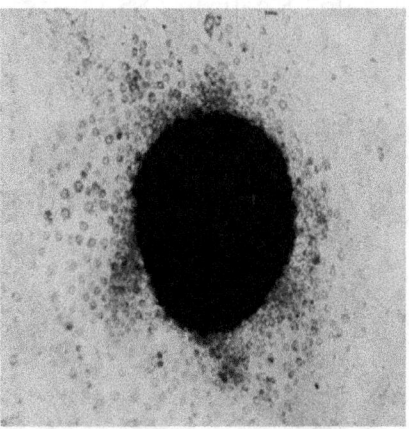

Alkaline Phosphatase staining of
buffalo ESC colony

Fig. 13.5: Alkaline phosphatase staining of buffalo embryonic stem cell colony

Stage Specific Embryonic Antigens (SSEAs)

The accessibility of molecules on the surface of cells make them exceptionally convenient markers for characterizing cell types, often recognized as antigen by specific antibodies. Cell surface antigens provide an invaluable tool for analyzing and sorting cells that have particular characteristics within specific contexts. A number of surface markers have been used for characterization of ES cells (Table 13.1). Stage specific embryonic antigens (SSEAs) which include SSEA-1, SSEA-3 and SSEA-4, are developmentally regulated during early embryogenesis and are widely used as markers to monitor pluripotency of ES cells. These SSEAs are various types of glycoproteins or glycolipids. The presence of SSEA-1 has not been documented in human embryo/ embryonic stem cell; however, pig, canine, chicken and sheep ES cells express SSEA-1. Conflicting reports are available in bovine for presence of SSEA-1 in ES cells, with some authors reporting its presence in ES cell-like cells but others not finding its expression. In our laboratory, we were able to detect its presence in buffalo embryonic stem cell lines. The other two most widely studied ES cell pluripotency surface markers are SSEA-3 and SSEA-4. They are related to globoseries cell surface glycolipids that were first used to delineate embryological changes in the developing mouse embryo. These markers were originally identified on human EC (Embryonic carcinimal) or ES cells and from primate ES cells like rhesus monkey. In mouse ES cells, SSEA-3 and SSEA-4 are expressed in 2 to 8-cell and morula stages of preimplantation embryos and are also found on unfertilized oocytes; however, there is a loss of expression in ICM of mouse embryos. In contrast, in human embryos, there is no expression of SSEA-3 or SSEA-4 at 2 to 8-cell or morula stages; however, these are expressed on ICM of human blastocysts and on isolated human ES cells. Whereas SSEA-3 and SSEA-4 are expressed in sheep and bovine, buffalo ES cell-like cells also show expression of SSEA-3 and SSEA-4. However, our group has observed lack of expression of SSEA-3 in buffalo ES cells but SSEA-4 was positively expressed in all our cell lines (Fig. 13.6).

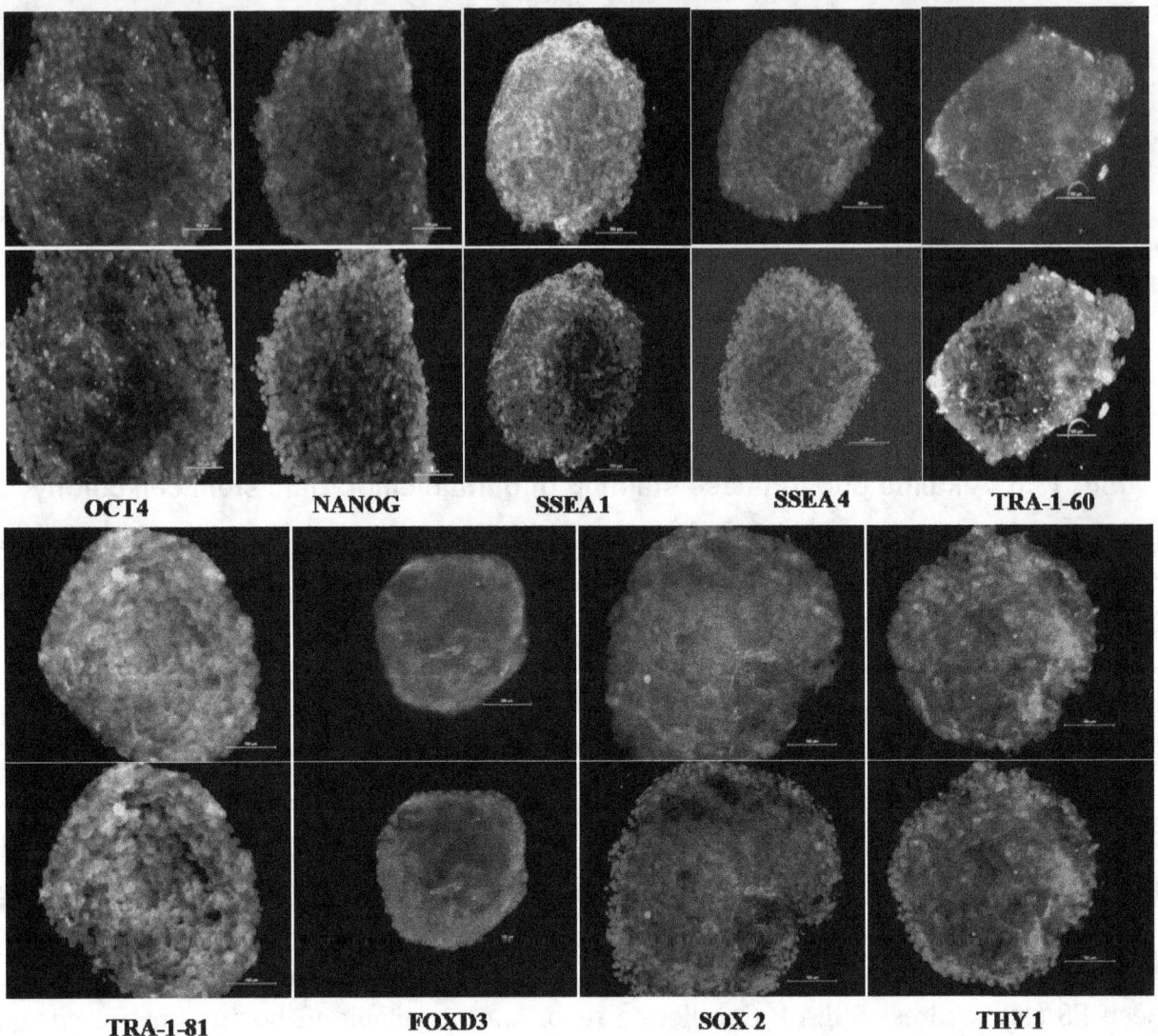

OCT4	**NANOG**	**SSEA 1**	**SSEA 4**	**TRA-1-60**
TRA-1-81	**FOXD3**	**SOX 2**	**THY 1**	

Fig. 13.6: Expression of embryonic stem cell markers by buffalo embryonic stem cell colonies upon immunocytochemistry

Tumour Rejection Antigens

Another class of surface markers for pluripotent cells is that of tumour rejection antigens (TRA) series. TRA-1-60 reacts with a sialidase-sensitive epitope while TRA-1-81 reacts with an unknown epitope of the same molecule. ES cells from primates like rhesus monkey have shown to express TRA-1-60 and TRA-1-81. The expression of TRA-1-60 and TRA-1-81 has also been reported in buffalo ES cell-like cells. We have also detected their expression in our ES cell lines (Fig. 13.6).

d) Transcription Factors as markers of pluripotency

A number of transcription factors play critical role in maintaining stem cell self-renewal, and their expression is being used to characterize ES cells in different species. Among them Oct4, Sox2, Nanog, Foxd3 and Rex1 are thought to be central to the transcriptional regulatory hierarchy that specifies ES cell identity because of their unique expression patterns and essential role during early development.

Table 13.1: Comparison of stem cell markers of embryonic stem cells of different species

Marker	Bovine	Buffalo	Murine	Human
Alkaline phosphatase	+	+	+	Variable
SSEA1	Variable	Variable	+	-
SSEA3	Variable	Variable	-	+
SSEA4	Variable	+	-	+
TRA-1-60	-	+	-	+

Oct4 is a transcription factor belonging to the class IV of POU family factors and is expressed by all pluripotent cells during embryogenesis. Oct4 has also been established as a marker for human pluripotent ES cells. However, in bovine cells, the usefulness of Oct4 has been questioned by identification of a bovine Oct4 pseudogene. Expression of Oct4 in undifferentiated pluripotent cells has also been shown in various other species like canine, goat and buffalo. We have detected the consistent presence of Oct4 in our ES cell lines.

Nanog is a homeobox-containing transcription factor with an essential function in maintaining pluripotency of ICM cells. Furthermore, over expression of Nanog is capable of maintaining the pluripotency and self-renewal characteristics of ES cells under the conditions where normally the cells would be exposed to differentiation-inducing culture conditions. Nanog seems to be one out of the several factors that are expressed in pluripotent cells and are down regulated at the onset of differentiation. Nanog mRNA was detected in the ICM but not in the TE of expanded goat blastocysts; a pattern that follows the expression observed in mice. We observed expression of this transcription factor throughout our established cell lines. We have likewise detected the expression of other transcription factors like Rex1, c-Myc, Sox2 in our established ES cell cultures both *via* immunocytochemistry as well as by RT-PCR.

e) Differentiation

The most definitive test of pluripotency is the formation of chimeras in mice, in which ES cells are injected into the blastocyst and contribution of the ES cells to the resulting chimera is assessed to determine the differentiation capacity of the injected cells. ES cells differentiate spontaneously *in vitro* when grown in absence of appropriate feeder cells. Under appropriate conditions, such as suspension culture, embryiod bodies (EBs) are formed in almost all species like canine, sheep, bovine and buffalo with region specific endoderm, mesoderm and ectoderm differentiation. We have proven the capability of our cell lines to form EBs with all the germ layers (Fig. 13.7). ES cells may be directed into lineage of interest by supplementing various growth factors or their antagonists into culture media. These growth factors or stimulating agents allow directed differentiation of ES cells towards a particular cell lineage or cell type. The differentiated cells should be identified with the help of various markers, which are highly expressed in these cells.

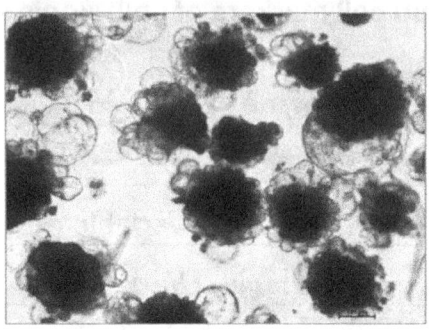

Day 14 EBs formed in suspension culture

Fig. 13.7: Embryoid bodies formed in static suspension cultures upon spontaneous differentiation of buffalo embryonic stem cells

Molecular mechanisms for pluripotency

Signaling molecules, transcription factors, cell cycle regulators and epigenetic modifications regulate intricate molecular mechanisms of maintaining pluripotency of embryonic stem cells among wide range of mammalian species. Embryonic stem cells require extrinsic factors to maintain their pluripotency in culture. These extrinsic growth factors act on different signaling pathways to regulate intrinsic transcription factor networks to sustain embryonic stem cells in an undifferentiated state. The signaling mechanisms regulating expression of pluripotency factors, Pou5F1, Nanog, Sox2 and Klf4 significantly differ in the context of species specification. The main extrinsic factors include Leukemia inhibitory factor (LIF), basic Fibroblast growth factor (bFGF) and Bone morphogenic protein 4 (BMP4) etc. These factors, when supplemented in embryonic stem cell culture, induce signals that are transmitted through intracellular components and regulate the expression of pluripotency factors. Intrinsic factors like ERK are present within the cell and in their active form generally induce the differentiation of mouse embryonic stem cells. Hence, inhibition of this signaling pathway maintains pluripotency of mouse embryonic stem cells.

Intrinsic factors in regulation of pluripotency

ERK signaling

All embryonic stem cells have an intrinsic property to differentiate spontaneously. Thus, the inhibition of differentiation should maintain the self-renewal character of embryonic stem cells. ERK1/2 signaling autoinductive stimulation by FGF or FGF4 is implicated in inducing differentiation of mouse embryonic stem cells. The suppression of FGF4/ERK signaling pathway has been shown to promote self renewal of mouse embryonic stem cells. However, complete inhibition of FGF/ERK signaling results in degeneration of mouse embryonic stem cells in culture. So to maintain metabolic activity, biosynthetic capacity and overall viability of these cells, it was felt important to inhibit Glycogen synthase kinase 3 or GSK3 signaling. This has been achieved by addition of CHIR99021 (CH) to stem cell culture. Thus, dual inhibition of ERK and GSK3 signaling by PD0325901 (PD) and CH, respectively, restored the pluripotency in mouse as well as rat embryonic stem cells in feeder free condition. It has been established that for maintenance of pluripotency and self-renewability, embryonic stem cells require a complex network of transcription factors namely POU5F1, SOX2, NANOG, OCT4 an KLF4. It has been shown that ERK1/2 binds at the C-terminal domain of KLF4 and phosphorylates it at Ser123 residue in mouse embryonic stem cells, thereby down regulating the transcriptional activity of KLF4 which in turn induces differentiation. Thus, inhibition of ERK signaling enhances KLF4 activity and in turn maintains the undifferentiated state of mouse embryonic stem cells. This is contrary to human and buffalo embryonic stem cells, where downstream of FGF signaling, activation of ERK signaling is associated with maintenance of pluripotency.

PKC signaling

Inhibition of protein kinase C (PKC) isoforms has been shown to maintain pluripotency in mouse embryonic stem cells without activation of STAT3 or inhibition of ERK/GSK3 signaling pathways. Atypical PKC isoforms, PKC function is involved in activation of Nfkb pathway during mouse embryonic stem cell differentiation. On inhibition of PKC signaling by the pharmacological drug, G06983, down regulation of Nfkb target genes was observed. Thus, it has been implicated that an appropriate activity of PKC-signaling directly balances the self renewal and differentiation of mouse as well as rat embryonic stem cells.

Aurora Kinase A/p53 signaling

It has been shown that loss of Aurora kinase A or Aurka is coupled to stimulation of p53 activity and induction of differentiation of mouse embryonic stem cells towards ectoderm and mesoderm lineage. Thus Aurka mediated p53 phosphorylation is essential for maintaining self-renewal and pluripotency in mouse embryonic stem cells. But it has not been established whether the function of Aurka is conserved among the species or not.

Extrinsic factors regulating pluripotency

LIF and JAK/STAT3 signaling

Mouse embryonic stem cells were originally cultured on feeder layers derived from mouse embryonic fibroblasts (MEF). It was later found that Leukemia inhibitory factor (LIF), a member of Interleukin-6 cytokines, is produced by these cells which acts as a key factor to maintain pluripotency of stem cells. Mouse embryonic stem cells are now cultured successfully *in vitro* in an undifferentiated state in presence of LIF and serum. LIF on binding to its receptor, recruits gp130 to form a heterodimer which subsequently activates Janus kinase (JAK), through transphosphorylation. Activated JAK then phosphorylates gp130, creating a docking site to bind to SH2 domain of Signal Transducers and Activators of Transcription 3 (STAT3). JAK phosphorylates the recruited STAT3, which forms a homodimer and subsequently translocates into nucleus, where it binds to gene enhancers to regulate target gene expression. Phosphorylated STAT3 essentially maintains the c-MYC level and induces KLF4 expression in mouse embryonic stem cells, which are essential transcription factors, maintaining pluripotency. STAT3 also induces Pramel7 activity which blocks phosphorylation of ERK. LIF receptor and gp130 are also expressed in human embryonic stem cells, and human LIF can induce STAT3 phosphorylation and nuclear translocation in human embryonic stem cells. However, human LIF is unable to maintain the pluripotent state of human embryonic stem cells, suggesting that mouse and human embryonic stem cells require distinct signaling mechanisms to govern their pluripotency. It has been found that STAT3 does not localize within nucleus of human embryonic stem cells and is incapable of activating the target genes. However, our studies revealed that buffalo embryonic stem cells are dependent upon LIF/JAK/STAT3 signaling for maintaining pluripotency and an undifferentiated state.

TGF-β signaling

TGF-β superfamily consists of more than 40 members, including TGF-β, Activin, Nodal and bone morphogenetic proteins (BMPs). The TGF-β members transduce signals by binding to heteromeric complexes of serine/threonine kinase receptors, type I and II, which subsequently activate intracellular Smad proteins. Smads 2 and 3 are specifically activated by activin, nodal and TGF-β ligands, whereas Smads 1, 5 and 8 are activated by BMP ligands. It has been shown that TGF-β related signaling pathways play complex roles in regulating the pluripotency and cell fate of embryonic stem cells.

BMP signaling pathway

It was observed that in absence of serum, LIF alone does not maintain the undifferentiated nature of mouse embryonic stem cells. But when these cells were cultured in presence of BMP4 and LIF, the undifferentiated status was retained. Bone morphogenetic protein (BMP) is a subset of the TGF-β superfamily. Dimeric BMP4 binds to type II receptors, BMPR2, and facilitates the assembly of receptor heteromers. They constitutively activate kinase domains of type II receptors and phosphorylate type I receptors (BMPR1). Activated type I receptors subsequently phosphorylate BMP-responsive SMAD1/5/8 which then form a complex with SMAD4 and translocate into nucleus to regulate target gene expression. In mouse embryonic stem cells, LIF can substitute mouse embryonic fibroblast feeder layers for maintaining pluripotency in presence of animal serum by activating the transcription factor

STAT3. However, in serum-free cultures, LIF is insufficient to block neural differentiation and maintain pluripotency. It has been reported that BMP4 is sufficient to replace serum to maintain pluripotency of mouse embryonic stem cells in presence of LIF. BMP has been sown to phosphorylate SMAD1/5 and activate inhibitors of differentiation (Id) genes, which block neural differentiation by antagonizing neurogenic transcription factors. However, in human embryonic stem cells BMP, in contrast to maintenance of pluripotency, has been shown to promote differentiation to trophoblasts. The inhibition of BMP signaling with BMP antagonist, Noggin, has been shown to sustain the undifferentiated state of human embryonic stem cells. In buffalo embryonic stem cells BMP4 has been shown to play a role in maintenance of pluripotency. We recently reported that BMP4 induces differentiation of buffalo embryonic stem cells towards germ lineage and inhibition of BMP signaling pathway by Noggin inhibits the germ line differentiation.

TGF-β/ Activin/Nodal signaling pathway

TGF-β and Nodal genes were shown to be highly expressed in undifferentiated human embryonic stem cells. Activin A, a member of the TGF-β superfamily, was found to be secreted by mouse embryonic fibroblast feeder cells. It has been shown that medium enriched with Activin A can replace mouse embryonic fibroblast feeder layers or feeder-conditioned medium to maintain human embryonic stem cells in an undifferentiated state. It has also been demonstrated that TGF-β/Activin/Nodal pathway is activated through the transcription factors Smad2/3 in undifferentiated human embryonic stem cells. It has also been demonstrated that TGF-β/Activin/Nodal signaling supports human embryonic stem cell self-renewal and pluripotency by inducing higher levels of pluripotent protein expression (Oct4 and Nanog), while inhibition of TGF-β/Activin/Nodal signaling with Lefty or Follistatin decreases expression of these pluripotent proteins in human embryonic stem cells. This further supports the role of TGF-β/Activin/Nodal pathway in maintenance of embryonic stem cell pluripotency and self renewal.

The signaling pathway has been shown to activate Smad2/3 which subsequently binds Nanog promoter in undifferentiated human embryonic stem cells to induce expression of Nanog, a pluripotent transcription factor. Nanog has also been shown to coordinate with Smad2 in a negative feedback loop to inhibit human embryonic stem cell differentiation. However, this pathway has been reported as not essential for pluripotency of mouse embryonic stem cells.

FGF/MEK signaling

Human embryonic stem cells were traditionally cultured in presence of fibroblast growth factors (FGFs) either on fibroblast feeder layers or in fibroblast-conditioned medium. It has been demonstrated that all the four FGF receptors (FGFR1/2/3/4) and the components (SOS1, PTPN11 and RAF1) of their downstream activation cascade are significantly up regulated in undifferentiated human embryonic stem cells. Withdrawal of FGFs or inhibition of

FGF signaling by a FGFR inhibitor, SU5402, rapidly induced human embryonic stem cell differentiation. FGFs signal by binding to FGF receptors (FGFRs) and activate multiple signaling cascades, including Mitogen-Activated Protein Kinases (MAPKs), Janus kinase/ signal transducer and activator of transcription (Jak/Stat). phosphatidylinositol 3-kinase (PI3K) and phosphoinositide phospholipase C (PLC) pathway. It has also been reported that FGF contributes to maintenance of human embryonic stem cells mainly through FGF/MEK pathway. It has been shown that FGF also induces feeder cells to secrete TGF-β1 and Insulin-like growth factor 2 (IGF2), which can subsequently promote undifferentiated state of human embryonic stem cells. However, FGF signaling is associated with induction of differentiation in mouse and rat embryonic stem cells. In buffalo embryonic stem cells, we have reported that addition of exogenous FGF2 plays a critical role in maintenance of pluripotent and self-renewal state, while culture of the stem cells in absence of bFGF induces differentiation, even in presence of buffalo fibroblast feeder layers. bFGF has also been reported to activate ERK1/2 signaling pathway which is responsible for maintaining pluripotency in human and buffalo embryonic stem cells. FGF signaling also inhibits spontaneous differentiation towards extra-embryonic lineage or neural induction in human embryonic stem cells. bFGF has further been reported to activate Nodal/Activin signaling upon activation of Activin A receptors, which further phosphorylate SMAD2/3 and form a complex with co-activator SMAD4. The SMAD4-complex induces expression of NANOG in the nucleus and supports pluripotency.

Wnt/β-catenin signaling

The canonical Wnt-pathway has been implicated in pluripotency of mouse embryonic stem cells. We have also recently showed the role of Wnt signaling in maintenance of buffalo embryonic stem cell pluripotency. It has been reported that Wnt induces association of the destruction complex composed of DVL1, AXIN, APC, GSK3 and phosphorylated β-catenin with phosphorylated LRP. This complex phosphorylates β-catenin but further blocks the ubiquitination by β-TrCP. As a result, newly synthesized β-catenin is accumulated in the cytosol and translocated into the nucleus. The β-catenin/CBP complex binds to its target, Tcf and Lef, and induces expression of Stat3 mRNA. Independent of Tcf/Lef binding, stabilized β-catenin enhances Pou5f1 activity for maintenance of pluripotency in mice ES cells. In human and buffalo embryonic stem cells, activation of Wnt/ β-catenin pathway with either WNT3A or BIO (a GSK3 inhibitor) maintains self-renewal of stem cells even under feeder-free conditions. Recently, Wnt/β-Catenin signaling has been reported to be inactive in self-renewal of human ES cells. It has been observed that during self-renewal in human ES cells, POU5F1 repress β-catenin signaling. It has been established that the enhanced expression of β-catenin is associated with induced differentiation in human ES cells. Intriguingly, the fate of differentiation of human ES cells corresponding to endogenous Wnt signaling is based on their heterogeneous culture. Human ES cells with a higher expression of Wnt predominantly differentiated towards endodermal and cardiac fates whereas those expressing lower level of Wnt, generated cells primarily from neuroectodermal lineage.

Pluripotency factor networks

All the external signaling events finally lead to distinct regulation of genes that result in pluripotent state. Thus, embryonic stem cell pluripotency is conferred by a unique transcriptional network composed of several transcription factors. The critical transcription factors identified as required for pluripotency include OCT4, SOX2, NANOG, FOXD3 and Id, etc.

OCT4 is a POU domain-containing transcription factor and is one of the first transcription factors identified as essential for both embryo development and pluripotency maintenance. In mouse and probably in most of the species, OCT4 expression is activated at 8-cell stage and is later restricted to the inner cell mass and germ cells. It has been found that decreased expression of OCT4 in embryonic stem cells leads to differentiation and loss of pluripotency. OCT4 has been found to regulate a broad range of target genes like Fgf4, Utf1, Opn, Rex1/Zfp42, Fbx15, Sox2 and Cdx2. Repression of Oct4 activity in embryonic stem cells up regulates Cdx2 expression, leading to embryonic stem cell differentiation into trophectoderm. OCT4 also activates downstream genes by binding to enhancers carrying the octamer-sox motif (Oct-Sox enhancer), for synergistic activation with Sox2. However, the precise level of OCT4 expression is very important in ES cell fate determination. It has been established that loss of Oct4 causes inappropriate differentiation of ES cells into trophectoderm, whereas its over expression results to differentiation into primitive ectoderm and mesoderm. In contrast to its target genes, little is known about Oct4 upstream regulators. Oct4 promoter contains conserved distal and proximal enhancers that can either repress or activate its expression depending on the binding factors occupying these sites.

SOX2 is an HMG-box transcription factor that is expressed in pluripotent cell lineages and nervous system. SOX2 forms complex with Oct4 protein to occupy Oct-Sox enhancers, which are found in regulatory regions of most of the genes that are specifically expressed in pluripotent stem cells such as Oct4, Sox2, Nanog, Utf1, Fgf4 and Fbx15 etc. Chromatin immunoprecipitation (ChIP) coupled with DNA microarray studies revealed co-occupancy within the same gene region by Oct4 and Sox2. It was further found that >90% of the promoter regions bound by Oct4 and Sox2 are also bound by Nanog, implicating to essential role of Nanog in embryonic stem cell pluripotency and self renewal. Nanog is also a homeobox-containing transcription factor that is specifically expressed in pluripotent embryonic stem cells. The essential role of Nanog in maintaining the pluripotency of ES cells is highlighted by the fact that Nanog-deficient ES cells are prone to differentiation, whereas forced expression of Nanog partially renders ES cells a self-renewal potential. It has been found that Nanog maintains ES cell pluripotency by: i) Downregulating downstream genes essential for cell differentiation such as Gata4 and Gata6; ii) Activating the expression of genes necessary for self-renewal such as Rex1 and Id. Nanog expression is in turn regulated by binding of Oct4/Sox2 and FoxD3 to the proximal region of Nanog promoter and support its expression, while TCF3 and p53 bind to the promoter and negatively regulate Nanog expression. LIF and BMP signaling and their downstream effectors of STAT3 are also involved in regulation of Nanog expression. Oct4, Nanog Sox2 and FoxD3 also bind to each other's promoter and support or limit its expression, forming an interconnected autoregulatory network to maintain ES cell pluripotency and self-renewal.

Adult stem cells

Adult stem cells are derived from different parts of the body and have different properties, depending on where they are derived from. They exist in several different tissues including bone marrow, blood and the brain. Some studies have suggested that adult stem cells are very versatile and can develop into many different cell types, while others have concluded that adult stem cells are only able to develop into a limited number of cell types related to the tissue from which the stem cells originally came from. The rate at which new cells are produced in the adult is a measure of how rapidly the cell population is turning over; and on this basis, tissues can be divided into three broad categories. In tissues with *static* cell populations, such as nerve and skeletal muscle, there is no cell division and most of the cells formed during development persist throughout adult life. In tissues containing *conditional renewal* populations, such as liver, there is generally little cell division, but in response to an appropriate stimulus most cells can divide to produce daughters of the same differentiated phenotype. Finally, tissues with *permanently renewing* populations, including blood, testis and stratified squamous epithelia, are characterized by rapid and continuous cell turnover in the adult. The terminally differentiated cells have a short lifespan and are replaced through proliferation of a distinct subpopulation of cells, known as stem cells. Stem cells that exist in postnatal tissues have been recognized since 1960s, with the first conceptual proof that blood or bone marrow contains cells which rescue humans and animals from bone marrow failures. It has been seen that even a single murine haematopoietic stem cell (HSC) can fully reconstitute all blood cell types following transplantation in lethally irradiated animals, and the progeny of such cells can reconstitute the haematopoietic system in secondary lethally irradiated recipients. HSCs, therefore, fulfill all the characteristics of stem cells. The term 'adult stem cell' is somewhat of a misnomer, because the cells are present even in infants, and similar cells also exist in the umbilical cord and placenta. More accurate terms have been proposed, such as tissue stem cells, somatic stem cells and postnatal stem cells. The scientific interest in adult stem cells rests on their ability to self-renew indefinitely, generate cell types of the organ from which they are derived, their potential to generate the entire organ and also to demonstrate a wide range of plasticity. Also, unlike embryonic stem cells, their use in not considered controversial as they are derived from adult tissues and not from embryos. These two factors and the associated hopes attract considerable interest in adult stem cell research.

Sources of adult stem cells

Adult stem cells can be obtained from bone marrow, hippocampus region of brain, cornea and retina of eyes, skeletal muscle, dental pulp, liver, skin, lining of gastrointestinal tract, pancreas, ovarian-epithelium and testis, etc.

Bone marrow stem cells

Bone marrow contains various discernable stem cell populations as discussed below.

Haematopoietic stem cells (HSCs)

HSCs are adult stem cells found mainly in bone marrow. They provide blood cells required for daily blood turnover and for fighting infections. The production of mature blood cells is a continuous process that is the result of proliferation and differentiation of stem cells, oligopotent progenitor cells and mature cells. A single stem cell has been found to be capable of more than 50 generations (doublings) and has the capacity to generate 10^{15} cells to support up to 60 years of life. They are commonly obtained from postnatal tissues such as bone marrow and blood, as well as from prenatal tissues, *viz.* umbilical cord, blood, bone marrow, liver, aorta, gonad, mesonephros region and yolk sac. Although the degree of self-renewal may differ for cells from ontogenically earlier or later HSCs, all HSCs regardless of the ontogeny have the same functional characteristics. HSCs are easy to obtain, as they can be either aspirated directly out of the bone marrow or stimulated to move into the peripheral blood stream, where they can be easily collected. HSCs are multipotent or pluripotent and capable to differentiate in various ways and thereby generate erythrocytes, granulocytes, monocytes, mast cells, lymphocytes and megakaryocytes. HSCs are few, normally fewer than one cell per 5×10^4 cells in bone marrow.

Ex vivo expansion of HSC

Dexter et al. were the first to reconstitute the *in vivo* haematopoietic microenvironment in *vitro* to culture progenitor stem cells. After one-and-a-half decades, Iscove et al. described a deliberate attempt for 8–12 fold expansion of murine pluripotent stem cells *in vitro* during four days of culture. In two decades of research, it is now evident that expansion of clonogenic progenitor cells is feasible in almost any type of *ex vivo* culture system. These studies demonstrated that over a reasonable time period, the number of progenitor cells expand several fold in culture. In some cases, progenitor cell numbers decline after attaining maximum density. Whether such decline results from adverse effects of *in vitro* expansion, or is due to natural senescence of more mature progenitor cells, is difficult to understand. Haematopoietic stem cell *ex vivo* cultures were initiated either with purified stem cells (CD34+) or mononuclear cells obtained from bone marrow, peripheral blood, or cord blood, in presence or absence of stroma cells, in stagnant or stirred medium, in serum-replete or serum-free medium, as well as in presence of different combinations of haematopoietic growth factors.

Phenotypic analysis of HSC

The most popular assay for HSC is its phenotypic analysis. Using flow cytometry analysis, it has been revealed that most active murine and human long term haematopoietic repopulating (LTR) cells are CD34−. However, for all practical purposes, CD34+ is considered as the most popular surface marker of HSCs. It is important to note that stem cells' population is not described based on one particular surface marker. For example, in case of murine bone marrow stem cells, the marker Sca-1 is not unique, as it is also expressed by some activated immune cells. At the same time, c-kit is a transmembrane tyrosine kinase receptor for stem cell factor (SCF)/steel factor (SF) and is considered to play an important role in early stages of haematopoiesis. Therefore, murine stem cells are confidently characterized as Sca-1$^+$c-kit$^+$ cells.

Mesenchymal stem cells (MSCs)

Friendenstein et al. first reported that bone marrow cells contain MSCs in addition to HSCs. These were initially isolated as the plastic adherent fraction of bone marrow. MSCs, also termed as marrow stromal cells, have been isolated from multiple tissues, the foremost being from bone marrow aspirates and from subcutaneous adipose tissue and fetal lung. Every blood vessel in the body has mesenchymal cells on the tissue side of the vessel, some of which are MSCs, although names like pericytes have been given to these multipotent cells. There is a consensus that MSCs do not express CD11b (an immune cell marker), glycophorin-A (an erythroid lineage marker), or CD45 (a marker of all haematopoietic cells). CD34, the primitive HSC marker, is rarely expressed in human MSCs, although it is positive in mice. CD31 (expressed on endothelial and haematopoietic cells) and CD117 (a haematopoietic stem/progenitor cell marker) are almost always absent from human and mouse MSCs. Stro-1 is by far the best-known MSC marker. CD106, or vascular cell adhesion molecule-1 (VCAM-1), is expressed on blood vessel endothelial and adjacent cells, consistent with a perivascular location of MSCs. CD106 singles out 1.4% of Stro-1-positive cells, increasing the CFU-F frequency to 1 in 3. These are all high Stro-1-expressing cells and are the only Stro-1-positive cells that form colonies and show stem cell characteristics such as multipotentiality, expression of telomerase and high proliferation *in vitro*. Taken together, these data suggest that Stro-1 and CD106 combine to make a good human MSC marker.

Multipotent adult progenitor cells (MAPCs)

MAPCs were isolated from human bone marrow in an attempt to isolate MSCs. They were believed to be mesodermal progenitor cells and were subsequently shown to differentiate into neurons and hepatocytes. Unlike most adult somatic cells, they appear to proliferate without senescence and have pluripotent differentiation ability both *in vitro* and *in vivo*. MAPCs can be cultured from human, mouse and rat bone marrows and also from mouse brain and muscle. The differentiation potential and expressed gene profile of MAPCs derived from these tissues appear to be highly similar. Unlike MSCs, MAPCs do not express major histocompatibility (MHC) class I antigens, are CD105 and SH2 (endoglin) negative and do not express or express only low levels of CD44 antigen. Unlike HSCs, MAPCs do not express CD45,

CD34 and c-Kit antigens but express Thy1, AC133 (Human MAPCs) and Sca1 (mouse), albeit at lower levels.

Marrow-isolated adult multilineage inducible (MIAMI) pluripotent stem cells

Another population of pluripotent cells from human bone marrow called marrow-isolated adult multilineage inducible (MIAMI) cells has been reported. This population was obtained by plating whole bone marrow cells initially in media containing 5% fetal bovine serum (FBS) and subsequently maintained in media containing 2% FBS in fibronectin-coated dishes. These cultures were maintained in hypoxic conditions (3% oxygen) at cell densities between 1300–1400 cells/sq cm. MAIMI cells express telomerase and transcription factors Oct4 and Rex-1 and were expanded for more than 50 population doublings. These were shown to differentiate into mesenchymal lineages as well as neural and pancreatic lineages.

Peripheral blood stem cells

There is abundant evidence that bone marrow stem cells can leave the marrow and enter into circulation. The specific mobilization of bone marrow stem cells is used to harvest stem cells more easily for various bone marrow stem cell treatments. Mobilized stem cells in peripheral blood have been administered intravenously in a rat model of stroke, ameliorating some of behavioral deficits associated with damaged neural tissue and leading to a proposal that stem cell mobilization in patients might be used as a treatment for stroke in humans. Mobilized stem cells have also been used in cardiac regeneration in mice. Two recent studies have found that human peripheral blood stem cells exhibiting pluripotent properties can be isolated from immobilized human blood. One study showed that the isolated cells were adherent, similar to marrow mesenchymal cells and could be induced to differentiate into cells from all three primary germ layers, including macrophages, T-lymphocytes, epithelial cells, neuronal cells and liver cells. Other study showed that induction of the peripheral blood stem cells could produce hematopoietic, neuronal or cardiac cells in culture. In the latter study, undifferentiated stem cells were negative for both major histocompatability antigens (MHC) I and II, expressed high levels of the Oct4 gene (usually associated with pluripotent capacity in other stem cells), and formed embryoid body structures in culture.

Spermatogonial stem cells (SSC)

Spermatogonial stem cells (SSC) belong to male germ line, which includes several families of undifferentiated cells with high developmental potential. These cells are biotechnologically important because they are the only cells among adult stem cells that are capable of transmitting their genetic information to future generations. This fact offers possibilities for SSC manipulation outside of the body in order to transfer relevant genes across herds. This is particularly important in cattle which have a long generation interval. These cells lie at the base of the seminiferous epithelium and coexist with their differentiating daughter cells in a space called the basal compartment, which is created by the surrounding sertoli cells. Proliferating cells committed to differentiate, which are located in that space, represent the transit amplifying compartment in the spermatogenic process where sertoli cells provide the niche for the process. SSCs and their initial differentiating daughter cells are extremely difficult to distinguish from a morphological point of view. Both cell types constitute a group of cells collectively called type A spermatogonia. Therefore, SSCs represent a subpopulation of type A spermatogonia group. During the formation of differentiating type A spermatogonia from a mother SSC, cytokinesis is not complete; so differentiating type A spermatogonia remain interconnected by intercellular bridges. In fact, in the adult testis, the first visible sign of the choice of SSC towards a differentiation fate is the formation of two daughter cells interconnected by an intercellular bridge.

Neuronal stem cells

Neuronal stem cells have been isolated from various regions of the brain, including more accessible olfactory bulb as well as spinal cord, and can even be recovered from cadavers soon after death. Evidence now exists that neuronal stem cells can produce not only neuronal cells but also other tissues, including blood and muscle. Animal studies have shown that adult neural stem cells can participate in repair of damage after stroke; either *via* endogenous neuronal precursors or transplanted neural stem cells. Endogenous neurons and astrocytes may also secrete growth factors to induce differentiation of endogenous precursors. It is also suggested that neural stem cells/neural progenitor cells may show low immunogenicity, being immune-privileged, on transplant. This raises the possibility for the use of donor neural stem cells to treat degenerative brain conditions. Using experimentally lesioned animals as models of Parkinson disease, human neural stem cells have been observed to integrate and survive for extended periods of time.

hNT cells

Embryonal carcinoma (EC) cells can be derived from teratocarcinomas of adult patients. These show multipotent differentiation abilities in culture. From one such isolation, a 'tamed' (non-tumorigenic) line of cells with neuronal generating capacity has been developed, termed hNT (NT-2) cells. Because of their capacity to generate neuronal cells, these cells have been studied for possible application in regeneration of neuronal tissues. hNT neurons show ability to generate dopaminergic neurons and have shown some benefit of transplantation in animal models of amyotrophic lateral sclerosis (ALS, Lou Gehrig disease). Early clinical trials using hNT neurons transplanted into stroke patients have shown initial positive results.

Muscle stem cells

Muscle contains satellite cells that normally participate in replacement of myoblasts and myofibres. There are also indications that muscle may additionally harbor other stem cells, either as haematopoietic migrants from bone marrow and peripheral blood, or as intrinsic stem cells of muscle tissue. Muscle appears to contain a side population of stem cells, as seen in bone marrow and liver, with the ability to regenerate muscle tissue. Muscle derived stem cells have been clonally isolated and used to enhance muscle and bone regeneration in animals. An isolated population of muscle-derived stem cells has also been shown to participate in muscle regeneration in a mouse model of muscular dystrophy. Stimulation of muscle regeneration from muscle-derived stem cells, as observed in other tissues, is greatly increased after injury of the tissue. An interesting use of muscle-derived stem cells has been regeneration and strengthening of bladder in a rat model of incontinence. Because of the similar nature of muscle cells between skeletal and heart muscle, muscle-derived stem cells have also been proposed for use in repairing cardiac damage, with evidence that mechanical beating is necessary for full differentiation of skeletal muscle stem cells into cardiomyocytes. Skeletal muscle cells have been used for clinical application to repair cardiac damage in a patient with positive results.

Liver stem cells

The liver is known to have a very high capacity of regeneration. The role of liver stem cells in regeneration has been controversial, but it is now accepted that these cells are important for the repair of specific types of liver injury. Liver stem cells might include: i) Cells responsible for normal tissue turnover; ii) Cells that give rise to regeneration after partial hepatectomy; iii) Cells responsible for progenitor-dependent regeneration; iv) Transplantable liver repopulating cells; v) Cells that produce hepatocyte and bile duct epithelial phenotypes *in vitro*. Human liver stem cells (HLSCs) express mesenchymal stem cell markers, CD29, CD73, CD44, and CD90, but not haematopoietic stem cell markers like CD34, CD45, CD117 and CD133. HLSCs are also positive for Vimentin and Nestin stem cell markers. The absence of staining for cytokeratin-19, CD117 and CD34 indicates that HLSCs are not oval stem cells. In addition, HLSCs express albumin, α-foetoprotein and in a small percentage of cells cytokeratin-8 and cytokeratin-18, indicating a partial commitment to hepatic cells. HLSCs differentiate to mature hepatocytes when cultured in presence of hepatocyte growth factor and fibroblast growth factor 4, as indicated by the expression of functional cytochrome P450, albumin and urea production. Under these conditions, HLSCs down regulate α-foetoprotein but express cytokeratin-8 and cytokeratin-18. HLSCs are also able to undergo oesteogenic and endothelial differentiation when cultured in appropriate differentiation media, but they do not undergo lipogenic differentiation. Moreover, HLSCs differentiate in insulin-producing islet-like structures. *In vivo*, HLSCs contribute to regeneration of liver parenchyma in severe-combined immunodeficient mice.

Hair follicle stem cells

The bulge area at the junction of the arrectores pili muscle to the hair follicle sheath has been shown to host the skin stem cells with maximum span of developmental potential.

These cells are maintained by signalling in concert with niche cells. Signals include paracrine, e.g. sonic hedgehog, autocrine and juxtacrine signals.

Mammary stem cells

Mammary stem cells provide the source of cells for growth of the mammary gland during puberty and gestation and play an important role in carcinogenesis of the breast. Mammary stem cells have been isolated from human and mouse tissue as well as from cell lines derived from the mammary gland. Such cells can give rise to both the luminal and myoepithelial cell types of the gland, and have been shown to have the ability to regenerate the entire organ in mice.

Induced pluripotent stem cells (iPSC)

The successful reprogramming of differentiated human somatic cells into a pluripotent state would allow creation of patient- and disease-specific stem cells. It was this idea that led Shinya Yamanaka and his team to successfully reprogram differentiated somatic cells into stem cells and win the 2012 Nobel Prize for physiology or medicine for the same. They reported generation of induced pluripotent stem cells (iPSCs) from somatic cells of mice (2006) as well as humans (2007) by transduction of four defined transcription factors viz, Oct3/4, Sox2, Klf4 and c-Myc. These iPSCs were similar in morphology, proliferation, surface antigens, gene expression, epigenetic status of pluripotent cell specific genes and telomerase activity to ESCs. iPSCs are an important advance in stem cell research, as they may allow researchers to obtain pluripotent stem cells, which are important in research and potentially have therapeutic uses, without the controversial use of embryos. Because iPSCs are developed from the patient's own somatic cells, it is believed that treatment of iPSCs would avoid any immunogenic responses, nevertheless further studies on the subject are essential owing to the retroviral methods of transduction that is used for generation of these cells. However, it has been demonstrated that the generation of iPS cells is possible without any genetic alteration of the adult cell by repeated treatment of the cells with certain proteins channeled into the cells via poly-arginine anchors that has been found to be sufficient to induce pluripotency and the cells produced in this manner are termed as piPSCs (protein-induced pluripotent stem cells.

Significance of farm animal stem cell research

Applications in regenerative medicine

Regenerative medicine is regarded as the future healthcare for it holds the potential for extending the reach of treatment modalities for individuals across diseases and lifespan. This aspect of medicine is believed to provide innovative solutions to complications ranging from congenital diseases and trauma to degenerative conditions. Stem cells, whether adult, embryonic or induced pluripotent, provide multipurpose and excellent research and clinical tools to understand and model diseases, develop and screen candidate drugs, and deliver cell-replacement therapies to support regenerative medicine. Stem cells are currently used to test/screen drugs or as the study material to identify molecules or genes implicated in regeneration. Reprogramming technology to produce iPSCs offers the potential to treat diseases like Alzheimer's, Parkinson's, cardiovascular, diabetes, and amyotrophic lateral sclerosis. In theory, easily-accessible cell types (such as skin fibroblasts) could be biopsied from a patient and reprogrammed, effectively recapitulating the patient's disease in a culture dish. Such cells could then serve as the basis for autologous cell replacement therapy. Since, cells originate within the patient, immune rejection of the differentiated derivatives would be minimized. As a result, the need for immunosuppressive drugs to accompany the cell transplant would be lessened and perhaps eliminated altogether. In addition, the reprogrammed cells could be directed to produce the cell types that are compromised or destroyed by the disease in question. Stem cell therapies are trialed in a number of medical complications to restore the function of a damaged tissue or organ like transplantation of patient's own stem cells (e.g. from bone marrow) to restore the cardiac damage and regenerate cardiac myocytes. The results have been mixed, though recent trials have shown a small but significant improvement in heart function. In another example fetal brain stem cells have been trialed as a treatment for Parkinson's disease, though with mixed results. The ethical issues associated with such cells, as well as their scarcity, suggest that other sources may have better long-term prospects. Stem cell therapy is seen as a highly promising area for type 1 diabetes, caused by autoimmune destruction of insulin-producing β-cells in the pancreas. Various approaches are being tested, including stored cord blood cells, human embryonic stem cells and mesenchymal stem cells. The latter are highly promising as they also seem to modulate the immune response to the beta cells. The hope is that mesenchymal stem cells will be grown outside the body, stimulated to develop into beta cells, and then returned to the patient to rejuvenate the impaired function. The liver is one of the few organs that can regenerate on its own accord as a subpopulation of cells in adults can divide and provide a new population of liver cells. Perhaps surprisingly though, progress towards artificial regeneration, either by stimulation of the liver cells or by transplantation of adult or embryonic stem cell-derived liver cells has been slow and remains highly experimental. Osteoporosis, a common problem in old age, could be countered by increasing the numbers of bone-forming osteoblasts which are formed from mesenchymal stem cells in the bone marrow. A slightly different but promising approach is to see stem cells as promoting regeneration rather than being the source of new building blocks themselves. The factors released by transplanted stem cells may encourage the growth and differentiation of stem cells already present in the damaged organ. This rationale has underpinned small-scale trials of mesenchymal stem cells in amyotrophic lateral sclerosis. There are encouraging signs that these cells can form a protective environment for motor neurons and slow down the loss of cells and development of symptoms. The applications of stem cell therapy, both in native state as well as genetically modified, is ever increasing encompassing almost every complications and it is beyond my scope to review all of them here. So much are the advances exciting that a dilemma exists as to whether continue research to learn more about stem cells and their properties is required or to go straight to the patients with the miraculous cells in hand.

Cloning Livestock – The Potential of ES cells

The first successful cloning of adult mammals (Wilmut et al. 1997) helped to overturn previous conceptions regarding the restricted developmental plasticity of somatic cells, and is raising exciting prospects for human regenerative medicine. In addition, somatic cloning, coupled with genetic modification, will better enable production of human therapeutics as well as xenograft tissues and organs from livestock. Opportunities for animal cloning and transgenesis in agriculture are more challenging than biomedical applications because they require greater biological efficiency and at reduced cost to be economically viable. For the production of transgenic livestock, embryonic stem cells might be beneficial than somatic cells, because they are more amenable to precise genetic modifications and result in higher cloning efficiencies than somatic cells in the mouse. Even with the most streamlined techniques, only 9% of zona-free somatic-cell cloned bovine embryos transferred to recipients result in viable calves at weaning. This level of cloning efficiency is considerably lower than the 40% of bovine IVF embryos that typically develop into healthy calves. The continuous loss of clones throughout pregnancy and high mortality during the perinatal period raise serious animal welfare concerns, especially for cattle, in which at least 60% of initiated pregnancies suffer severe placental abnormalities. These losses have mostly been attributed to faulty epigenetic reprogramming of the donor cell genome, resulting in major dysregulation of gene expression, particularly in the placenta with long-lasting effects into adulthood in some surviving somatic cell clones. Evidence from mouse cloning suggests that choosing particular donor cell types for NT (Nuclear transfer) could overcome many of these cloning-specific problems. Although no well-controlled studies have directly compared cell types within one cellular lineage of the same genotype, it is nonetheless clear that the degree of donor cell differentiation affects cloning efficiency. A higher proportion of transferred cloned embryos reconstructed from less-differentiated embryonic blastomeres or (murine) embryonic stem (ES) cells (pluripotent cells derived from embryonic blastomeres) result in viable offspring when compared with somatic cell clones. Cloning efficiencies with blastomere donors are approximately one order of magnitude higher than with somatic cells (36% versus 0.6% in mice and 28% versus 5% in cattle). Although abnormal phenotypes are still observed with blastomere cloning, their incidence and severity are greatly reduced. Cloning efficiency with F1 mouse ES cells is similarly remarkable, with 10–20% of transferred embryos reaching adulthood, compared with only 1–3% in the case of commonly used somatic donor cells (fibroblasts and cumulus cells) and less than 0.03% for terminally differentiated cell types (B or T lymphocytes). The candidate gene expression profiling shows that all ES cell-derived blastocysts faithfully express key embryonic genes such as Oct4 and related genes present in the pluripotent cells of the early embryo, whereas 38% of cumulus-cloned blastocysts failed to re-activate these genes. Although such detailed molecular analyses have not yet been performed with blastomere-derived cloned embryos, it is likely that blastomeres also retain an epigenotype compatible with early embryonic development without the need for extensive reprogramming. We , at Embryo Biotechnology Lab of National Diary Research Center, have been successful in cloning buffalo calves from a wide variety of cells ranging from buffalo fetal ear fibroblasts (Garima), adult ear fibroblasts (Purnima and Shresth), seminal epithelial cells (Swarn) to embryonic stem cells (Garima II). Garima II achived normal puberty and delivered a normal female buffalo calf (Mahima) following artificial insemination. The mother-daughter is perfectly normal, unlike other clones which developed some abnormalities, and are an example and inspiration within themselves.

Challenges in stem cell research: Looking ahead to 2017

The stem cell research field has reached a critical juncture in 2013 where we found ourselves buoyed by building momentum for both transformative basic science discoveries and clinical transformation of the stem cells. The overall prospect of novel stem cell based therapies becoming a reality comes with its own challenges apart from the technical ones, like stem cell tourism, formal physician training in stem cells, regulatory compliance balanced with innovation, savvy educational outreach and the risk of unproven therapies. Stem cell tourism is a small but growing part of the thriving global medical tourism marketplace where a patient travels to other country for getting stem cell therapy which is per se banned in his own country. Though much stem cell research remains at the experimental stage, with clinical trials still uncommon, yet there are over 700 clinics estimated to be operating in mostly developing countries like Costa Rica, Argentina, China, India and Russia, that are enticing many patients, mostly from industrialized countries. These stem cell tourists driven by desperation and hope, in turn continue to fuel the growth of such tourism. This '*magic cure by stem cells*' approach must be condemned under all circumstances. If there is no chance of improvement in the patient's condition, the 'therapy' is both unethical and scientifically and clinically unacceptable. The risks for adverse effects may be high; and it will not contribute to the development of clinically-established stem cell therapies, thus cutting the axe at its own feet. Prevention of stem cell tourism, together with a formal training of the stem cell based health practitioners curb on stem cell myths and misinformation propagated by TV shows, magazines, movies and other pop-culture avenues which remains the major challenges of the day. The stem cell community needs to take action in the coming time to use social media to promote evidence based stem cell medicine so as to take care of the hope and make the distinction clear between the 'hope or illusion' and the 'informed hope.' According to Knoepfler (2014), the key trends in stem cell field to watch for are:

1) Social media and pop culture influence
2) Formal physician training

3) Regulatory compliance balanced with regulation

4) Changes and possible receding of stem cell tourism rates

5) A more cooperative environment between scientists, physicians, patients, advocates, politicians, etc.

The potential of stem cell research for treatment has received praises from all over, from genuine idealism and optimism towards the practical ability of modern medicine. This however, warrants the need for more rigorous research and therapy based on evidence not on hope. Human beings would be always attracted towards the sciences which offer the hope of escaping diseases, death and decay, and within this backdrop stem cell science was and is seen with much hope and hype. It is this confluence of hope and hype that created demand for stem cell treatments that the science of clinical applications was not yet ready to accommodate. This is an effect which the stem cell advocates themselves contributed by rousing the public excitement, to the creation of offshore markets which are a havoc of the time. Thus, the need of the hour is to more vigorously criticize the charlatans who under the disguise of modern treatment either offer a placebo to the patients or do the harm in the hope of cure, depriving the people both economically as well as medically. More extensive research must be performed using stem cells from animals which are presumed to be closer to humans than mice and multiple animal models across species must be raised for a particular diseases for which stem cell therapy or intervention is sought, before the cure could be offered to the ailing humanity.

Further reading

Colman A and Kind A. (2000) Therapeutic cloning: concepts and practicalities. Trends in Biotechnology, 18: 192.

Evans MJ and Kaufman M. (1981). Establishment in culture of pluripotent cells from mouse embryos. Nature, 292: 154.

Evans MJ, Notarianni E, Laurie S and Moor RM. (1990). Derivation and preliminary characterization of pluripotent cell lines from porcine and bovine blastocysts. Theriogenology, 33:125–128.

Zandi M, Musharifa M, Syed MS, Ramakant K, Singh MK, Palta P, Singla S, Manik RS and Chauhan MS. (2013). WNT3A signalling pathway in buffalo (Bubalus bubalis) embryonic stem cells. Reproduction, Fertility and Development., dx.doi.org/10.1071/RD13084.

Shah SM, Saini N, Ashraf S, Zandi M, Manik RS, Singla SK, Palta P and Chauhan MS. (2015). Development, characterization and pluripotency analysis of buffalo (Bubalu bubalis) embryonic stem cell lines derived from *in vitro* fertilized, hand-guided cloned and parthenogenetic embryos. Cellular Reprogramming, 17: 306–322.

Odorico J, Kaufman D and Thomson J. (2001). Multilineage differentiation from human embryonic stem cell lines. Stem Cells, 19:193–204.

Stojkovic M, Lako M, Stojkovic P, Strachan T and Alison M. (2004). Derivation, growth and applications of human embryonic stem cells. Reproduction, 128:259–267.

Shah SM, Saini N, Syma A and Chauhan M. (2012). Bioinformatics in stem cell characterization. Indian Journal of Bioinformatics and Biotechnology 1:17–18.

Takashahi K, Koji T, Ohnuki M, Narita M, Ichisaka T, Tomoda K and Yamanaka S. (2007). Induction of pluripotent stem cells from adult human fibroblasts by defined factors. Cell, 131: 1–12.

Zhou H, Wu S and Joo J. (2009). Generation of Induced Pluripotent Stem Cells Using Recombinant Proteins. Stem Cells, 4: 381–4.

Dutta D. Signaling pathways dictating pluripotency in embryonic stem cells. (2013). International Journal of Developmental Biology, 57: 667–675.

Shah SM, Gigvijay K, Saini N, Aijaj S, Ashsraf S, Chauhan MS and Rana DS. (2012). Sources and culture strategies of adult stem cells. Current Medical Research and Practice, 2: 341–346.

Pan G and Thomson JA. (2007). Nanog and transcriptional networks in embryonic stem cell pluripotency. Cell Research, 17: 42–49.

Syed Mohmad Shah and Manmohan Singh Chauhan. (2015). Development of buffalo (*Bubalus bubalis*) embryonic stem cell lines from somatic cell nuclear transferred blastocysts. Stem cell Research, 15: 633–639.

Syed Mohmad Shah, et al. (2015). Cumulus cell conditioned medium supports embryonic stem cell differentiation to germ cell-like cells. Reproduction Fertility and Development 2015, doi.org/10.1071/RD15159.

Syed Mohmad Shah, et al. (2016). Testicular cell conditioned medium supports embryonic stem cell differentiation towards germ lineage and to spermatocyte- and oocyte-like cells. Theriogenology http://dx.doi.org/10.1016/j.theriogenology.2016.02.025.

Syed Mohmad Shah, et al. (2016). Spontaneous differentiation of buffalo (*Bubalus bubalis*) embryonic stem cells towards germ cell lineage. Journal of Stem Cell Research and Medicine 2016, doi:10.15761/JSCRM.1000103.

Chapter 14
Transgenic Animal Technology

Introduction

Ever since man learnt to domesticate animals and habituated them to live in his proximity, he has been and still is in search of a technique to propagate the useful and more important ones for his benefit and survival. The initial selection was predominantly done on the basis of phenotype and specific traits. An increasing knowledge in population genetics and statistics resulted to scientific animal breeding which incorporated one or more biotechnological procedures. The aim of all these techniques was efficient exploitation of the genetic potential of valuable sires and their propagation in a given population. With increasing understanding of the genetics and hereditary, was born a new technique of recombinant DNA technology which made possible to transfer a gene of interest from one organism (for example human) into another (for example bacteria). The transferred gene would be expressed resulting in production of a foreign protein in a distinct and unrelated species. This inevitably brought a revolution when human therapeutic proteins, enzymes and hormones were produced in bacteria and harvested in pure and active forms. However, reduced activity was found in such proteins produced in bacteria or other lower organism like fungi, insects etc., primarily owing to ontogenetic distance and poor or no post translational modifications in the proteins of interest. It was felt that mammalian specific post translational modifications could be achieved only when the proteins are produced in higher mammalian species, like farm animals. The production of such proteins in animal milk or urine would provide the natural and biological manufacturing unit for these proteins with no additional costs than rearing an animal. This led to the development of animal transgenic science and gradually to a full-fledged animal transgenic industry.

Transgenic animal

A transgenic animal is an animal whose genome has been altered by inclusion of foreign genetic material. The foreign genetic material is introduced through recombinant DNA technology, also known as genetic engineering. This technology/ engineering refers simply to a group of techniques used to cut apart and splice together pieces of DNA. The purpose of adding a new gene to an organism's genome is to have the organism produce a protein or set of proteins that it could not produce with its natural genome and without the addition of that specific gene. The added gene is usually from another species (Trans-genic) or sometimes from the same species (Cis-genic). The newly produced protein is usually of tremendous pharmacological or nutritive value. The gene of interest which is desired to be expressed is referred to as *transgene* or *insert*. The introduction of this foreign gene into animal cell or blastocyst is known as *transfection* (in contrast to *transformation* which refers to delivery into bacteria). The insert is not delivered as an isolated fragment of DNA but rather under a suitable promoter and with other necessary elements like insulator and terminator sequences. This is referred to as *gene* or *transgenic construct* and in some cases also contains intron sequences within the coding sequence of the desired protein. The gene construct is amplified and propagated before transfection in appropriate plasmids known as *cloning* (when only propagation is sought) and *expression* (when expression of the desired protein is sought) vectors. The vectors contain selectable markers (antibiotic resistance genes) which aid in selection of the transfected cells from non-transfected cells in culture in presence of a suitable selector (Genticin), origin of replication for independent replication of the plasmid/ vector, and other elements which vary depending on the type of the vector.

History of transgenesis

The essence of permanent evolution in living organisms was not perceived by humans until the invention of agriculture and breeding. The control of plant and animal reproduction made possible the empirical genetic selection which provided to human communities essentially all their food products, pets and ornamental plants. This led to generation of profoundly genetically modified organisms, some of which like carrots, tomatoes, silk worm, some dogs etc., got so dependent on human breeding that they were unable to survive without human assistance. With the re-discovery of Mendel's laws and increased understanding of genetic and molecular basis of trait transfer, inheritance and mutations, newer methods of breeding and genetic selection came into being. Initially most of the selection was based on spontaneous, thus random and unknown, mutations. Thus, it was necessary to increase the number of random mutations to enlarge the choice of genetically modified organisms corresponding to expectations of experimenters, farmers and breeders. This was achieved by using chemical mutagens and by generating multiple intra-and inter-species hybrids. One such impressive example was the creation of new cereal, *triticale*, from artificial crossing of wheat and rye. These methods, though beneficial, were imprecise and in few cases harmful as they induced multiple unknown mutations in addition to the expected ones. The discovery of DNA and genes, and further knowledge of gene structure and function opened wide avenues for research and biotechnological applications. Not only were the selection and mating more precise, accurate and predictive but the introduction of one trait from an organism into another was also aimed it. This eventually led to development of techniques and tools for genetic engineering and gene manipulation. The discovery of restriction enzymes which would cut DNA segment at a specific sequence was a landmark discovery in genetic engineering. This was further boosted by discovery of Ligases, enzymes which join together two DNA fragments irrespective of origin and relatedness. These two enzymes together with the understanding of gene expression led to development initially of genetic engineering and gradually of transgenic animal production.

The introduction of isolated genes into cells became a common practice in late 1970s, soon after the emergence of genetic engineering techniques, leading to production of the first chimeric mouse. A chimera is an animal that consists of two or more tissues that have different genetic compositions. The chimera mouse was constructed by introducing cells from one strain of mouse into embryos of another strain by microinjection into the blastocyst-stage embryo, a day 5 embryo containing 100–150 cells. The embryo was then implanted into a surrogate mother and allowed to grow into a chimeric adult mouse that exhibited the characteristics of both the strains. The next step in development of transgenic technology was use of retroviruses to deliver foreign DNA into embryos. Though successful, it lead to high degree of mosiacism, a condition in which only some of the cells in the tissues of an organism receive the genetic change, while other cells have the original genetic material without the desired addition. Another problem with this technology was interference of viral sequences with transgene expression. Using this retro-viral gene delivery technique, Gordon and Ruddle created the first transgenic mice in 1980–81. They coined the term "Transgenic" to describe the mice carrying new genes. The production of these transgenic mice led to the discovery of new techniques for creating transgenic animals.

With advancement of time and knowledge, embryonic stem cell mediated transgenic techniques were developed. The advantage with these cells was their pluripotent nature, implying that they could be grown into any of the cell types in the body. A host of other DNA delivery techniques were developed ranging from physical, chemical, viral to electrical modes. These will be discussed briefly in the following sections.

Development of a transgene construct

Although the genetic code is essentially the same for all organisms, the fine details of gene control differ. A gene from a bacterium will not often work correctly if it is introduced unmodified into an animal cell. Thus, it becomes essential to precisely construct a transgene. The simplest construct consists of the gene of interest plus some extra DNA that correctly controls the function of the gene in the new animal. The controlling DNA element is known as *promoter*. The selection of promoter depends on the level of expression desired, tissue in which the protein is finally expressed as well as the control on expression. A transgene also requires a *poly A* sequence for its correct function. In general, most of the constructs are constructed such that the gene/ cDNA of interest is placed under the control of a heterologous promoter, whose choice depends upon where and when the transgene is desired to be expressed. For a protein to be expressed, the cDNA of gene sequence must contain a translational start codon (ATG) with an upstream Kozak sequence (GCCGCC(G/A)NN) to provide for ribosomal recognition of mRNA start site and an in-frame translational stop codon (UGA, UAG, UAA) for translational termination. It has been proposed that inclusion of an intron at the 5' or

3' end of the transgene allows splicing of the transgene. Splicing generally results in more stable mRNAs, and efficient RNA translocation from nucleus to cytoplasm which typically leads to better transgene expression. Natural introns as well as artificial introns have been used. The eukaryotic stop signals that include poly (A) - addition sequence (AAUAAA) are positioned at the 3' end of protein translation sequence, followed by termination sequences. The termination sequences widely used include those for SV40, bovine growth hormone and human growth hormone. The enhancer and promoter together control the level and pattern of gene expression and are solely responsible for tissue-specific protein expression. Enhancer sequences are genetic control elements that act in position- and orientation-independent manners to control the level and pattern of gene expression. The transgene construct also includes insulator sequences or boundary elements at its 5'-end in order to insulate it from other control elements when it gets randomly integrated into the host cell genome. Most common insulator elements are 5'HS4 chicken beta-globin and mouse tyrosinase locus control region (LCR) insulator element. These have been shown to reduce the variability of transgene expression when introduced either 5,' or 5' and 3' relative to heterologous transgenes. The chicken lysozyme locus also has two scaffold/matrix-associated regions (S/MARs) surrounding the gene that has been shown to exhibit boundary-type functions in transgenic mice. A SINE B2 element of the family of short interspersed repetitive DNA elements has also been reported to function as an insulator at the mouse growth hormone locus.

Insertion of the transgene

A transgenic organism results from an inheritable genetic modification induced by the artificial transfer of an exogenous DNA fragment. This implies that the introduced transgene is integrated into a chromosome that could be transmitted to progeny as a host gene. For this purpose, the transgene has to be present in gametes of transgenic animal in order to be inherited by progeny. To achieve this, the foreign gene has to be introduced in embryo at first cell stage by a direct microinjection or *via* gametes containing the transgene. The foreign gene may also be introduced into somatic cells to develop a true transgenic cell line. The transgenic cells could then be used as donor cells in reproductive cloning to produce transgenic cloned embryos. A number of different methods have been successfully used for production of transgenic animals, some of which are a subject matter of this section.

1. Pronuclear microinjection

Pronuclear microinjection is a straightforward procedure which involves placement of DNA containing gene of interest into pronucleus of zygote, followed by transfer of zygotes to surrogate mother. This was the first technique used to generate transgenic mice by Gordon et. al. in 1980, and was subsequently extended to other animals. Pronuclear microinjection is the most popular and most successful technique used to create transgenic animals. The transgene construct is first prepared in a plasmid or cosmid and propagated suitably. Oocytes are harvested from superovulated females or aspirated from slaughter-house derived ovaries. They are fertilized *in vitro* and the newly fertilized eggs are harvested for pronuclear microinjection. Pronucleus refers to nucleus of a sperm cell (male) or egg cell (female) before they join to become fertilized zygote. The recombined vector (plasmid or cosmid) is splitted using restriction endonucleases and the gene construct is extracted, precipitated, washed and placed in an injection buffer. The extracted DNA solution is diluted so that one picoliter contains about 1000 copies of the gene construct. The equipments needed for microinjection include an inverted microscope, two micromanipulators and injection equipment, injection chamber and holding and injection pipettes. The injection pipette is filled with DNA solution and zygote is held with holding pipette. The injection pipette is introduced into the pronucleus, passing through zona pellucida, the cell membrane and the nuclear membrane. About 1–2 picoliters of DNA are injected into pronucleus. Most often male pronucleus is injected owing to its bigger size and clearer visibility. The injected zygotes are transferred, after brief *in vitro* culture, into the oviducts of synchronized recipient animals. After the birth of offspring, high molecular weight DNA is isolated from tissue of the offspring to confirm integration. The integration of injected DNA and number of copies can be determined by Southern blotting, Dot Blot hybridization, PCR or qPCR. Integration sites in chromosomes can be proved through hybridization of metaphase chromosomes using the injected gene as the probe. The success of pronuclear injection with respect to transgene integration ranges from around 3% in mice, rats and rabbits to only 1% for cattle, pigs and sheep. This also results in a high percentage of mosaics in which not all cells of the animal contain transgene. The time and cost of screening for germline transmission in mosaic animals prohibits generation of more transgenic animals through breeding. The procedure also leads to high variability in transgene expression between animals due to mosaicism, variable efficiency in transgene integration and chromosomal position effects that occur during random integration of transgene. It also allows for random addition of exogenous DNA rather than targeting to specific sites. Transgene insertion has also been found to alter the expression of endogenous genes at integration site. Transgenic animals so found positive for the transgene are raised and mated. Offspring from these matings are tested to discover whether the transgene has been passed on. In mating hemizygous (containing transgene on only one of the homologous chromosome) transgenic F1 *inter se* attempts are made to produce the homozygous progeny for the transgene.

2. Transposon mediated insertion

Transposons are short genomic DNA regions which are replicated and randomly integrated into the same genome. The number of a given transposon is thus increasing until the cell blocks this phenomenon to protect itself from a degradation of its genes. Foreign genes can be introduced into transposons *in vitro*. The recombinant transposons are then microinjected into zygote, where they become integrated with a yield of 1%. All the transgenic insects are being generated by using transposons as vectors. Transposons also proved efficient to generate transgenic fish, chicken and mammals. Transposons are efficient tools but they can harbor no more than 2–3 kb of foreign DNA.

3. Viral delivery

The efficiency of viral vectors to produce germline transgenic animals (capable of transmitting transgene to subsequent generations) has been much more restricted, despite of their extensive use in delivery of transgenes to somatic cells for gene therapy purposes. Both retrovirus (lentivirus) and replication defective adenovirus have been used as vectors to introduce exogenous DNA into animals. During a retroviral infection the genetic material is released into the cells as RNA, which is subsequently reverse-transcribed to DNA and integrated into the host genome by retroviral integrases. The only way of reaching of the retroviral pre-integration complexes and subsequently integrating into the host's chromatin is after nuclear membrane breakdown during mitosis. Thus, retroviral integration is restricted to dividing cells. Initial attempts to produce transgenics by retroviral infection of early embryos invariably resulted in genetic mosaics caused by multiple insertion sites of the transgene. There is a short interval of opportunity for the viral preintegration complexes to reach the embryonic chromatin during M-phase. This is the reason for delayed viral insertion and also in cell lineages with different insertion sites or no insertion site at all. A significant advancement of this technique has been achieved by exposing metaphase II (MII) oocytes to transgene containing retrovirus. The arrested oocytes have an advantage as they have already undergone nuclear envelope breakdown and remain at MII stage for a longer period of time as compared to M-phase of a somatic cell. This maximizes the probability of preintegration complexes gaining access to oocyte chromatin. Reverse transcribed gene transfer has also been used and this has proved to be much more efficient as far as integration was concerned. An alternative to retroviral delivery is adenoviral transmission. Adenovirus can infect a wide range of cell types, can accommodate large fragments of exogenous DNA (>20kb) and produce high viral titres. Despite these obvious advantages, adenoviral delivery is seldom preferred because of less effectiveness of gene transfer into gametes and embryos. Further studies are required to find an appropriate dose of infective particles that would render maximal integration frequency with acceptable toxicity and improvement in viral construct design.

4. Sperm mediated delivery

The ability of sperm cells to carry exogenous DNA into oocyte during fertilization was first reported by Brackett and coworkers in 1971. The technique however went in abyss for further 18 years till Lavitrano and coworkers reported the use of spermatozoa as DNA carriers to produce transgenic mice. Since then, transgene delivery by sperm cells has been used to produce transgenic animals in a wide variety of species like cattle, pig, rabbit, frog and zebrafish. It is perhaps the most straightforward approach envisioned to date for production of transgenic animals. The method involves incubation of spermatozoa with DNA containing the gene of interest followed by *in vivo* or *in vitro* insemination. The DNA binds to sperm plasma membrane through specific DNA-binding proteins and part of it (15–20%) is internalized by a mechanism probably mediated by CD4 molecules and carried into the oocyte upon fertilization. The methodology is appealing for production of transgenic animals due to its simplicity as it involves no embryo manipulation and possibility of achieving mass production of genetically modified animals through *in vivo* or *in vitro* insemination of many oocytes. This technique is however limited as it involves no targeted modification of homologous recombination as the integration process is random. However, in a recent report successful generation of transgenic pigs carrying a human gene (human decay accelerating factor) has been provided. By this method, the researchers (Lavitrano and coworkers) reported 80% transgene integration, and expression in 53% animals. This success represents an important step towards production of humanized pig organs and tissues for human transplantation. This method has been greatly improved upon by using ICSI (intracytoplasmic sperm injection), which consists of injecting the DNA-laden sperm into ooplasm. Sperm plasma membrane is initially damaged by freezing and thawing before incubation in presence of gene of interest. This method has proved efficient in mice and pigs and is believed to be extended to other animals also. Transposon use and ICSI may be combined in future to increase the yield of transgenesis.

5. Somatic cell nuclear transfer

Nuclear transfer, or cloning, is a technique that can be used to create a genetically identical copy, or a clone, of an animal. It involves transfer or placement of a donor nucleus into the cytoplasm of an enucleated MII oocyte. The donor cell could theoretically be any cell transfected with the transgene and showing its stable expression. Initially embryonic blastomers were used as donor cells but the process was hampered by limited number of cells available in an early embryo. Currently fetal and adult cells as well as embryonic stem cells have been used successfully to clone all the major livestock species including sheep, cattle, buffalo, goat and swine. This ability to use cells that can be cultured increases the number of cells available to clone, thereby facilitating the ability to make transgenic animals. The transgenes are introduced into cultured cells by physical/chemical methods like *Calcium mediated delivery, lipofection, necleofection, elctroporation* or by *viral transduction*, and the transfected cells are finally used as donor cells for SCNT. This methodology has facilitated the ability to make transgenic animals by circumventing most of the shortcomings of other transgenic techniques. One of the most important advantages is that the sex of the animal can be predetermined by choosing the donor animal (male or female) cell line. The use of cell culture to propagate donor cells also leads to large number of transgnic cells that can be stored and frozen for long periods of time. These transgenic donor cells can eventually give rise to numerous cloned transgenic animals. The transgene structure and expression can be tested by molecular techniques like PCR, Southern blot analysis, fluorescence *in situ* hybridization, western blot analysis etc. before initiating nuclear transfer and transferring the embryo to the recipient cow with a lengthy gestation time of 9 months. The proper use of SCNT ensures that 100% of animals produced are transgenic and that every cell of a cloned animal will have the transgene, thereby saving cost and time associated with recipient animal production. The ability to use a clonal population of transgenic cells guarantees the same transgene insertion site for each clone, thus decreasing animal to animal variation in transgene expression levels. Another advantage is that the transgene can be added to a specific genetic background. For example, a female that is above average in milk protein production may be used as the genetic background (donor somatic cells) in which the transgene is placed. SCNT also allows for targeted addition of DNA by homologous recombination which is vital in modulating specific gene expression as well as creating gene knockouts.

The success rate for somatic cell nuclear transfer averages 1–3% in most animals. Majority of embryos are lost during pregnancy with a 60% higher fetal loss between gestational days 35–60 when compared to embryos created through *in vitro* fertilization. In cloned cattle, there is higher perinatal loss than that observed in general population. These losses are due to host of complications like increased birth weight (large offspring syndrome), pulmonary abnormalities, respiratory problems, placental abnormalities and metabolic deficiencies. Most of these anomalies are believed to originate due to deficient or inadequate resetting of the developmental clock present in differentiated nucleus used as donor of the genetic material. The mechanism and factors affecting this process of nuclear reprogramming are not exactly known but a series of factors have been identified which increase reprogramming capability of the donor genome and promise for increased cloning efficiency. Reprogramming is loosely defined as a set of epigenetic changes required for a nucleus to change developmental fates. During SCNT, oocyte changes the fate of the donor nucleus from its original status to that of a zygotic nucleus. The reprogramming is not as exhaustive as occurs during natural or *in vitro* fertilization, resulting thereby in pregnancy losses. MII arrested oocytes are considered as the cytoplast of choice for nuclear transfer procedures. High levels of Maturation promoting factor (MPF) present in MII oocytes has been associated with successful nuclear reprogramming. MPF activity is maximal in both MI and MII oocytes but rate of blastocyst formation in embryos reconstituted with somatic cells and MI oocytes is significantly lower than that of embryos reconstituted with MII oocytes. Thus, MII oocytes rather than MI are considered more appropriate recipients for production of differentiated cell-derived cloned embryos. However, a high variability in quality of oocytes has been reported and has been proposed to be the main reason of abnormal epigenetic reprogramming. Though both *in vivo* and *in vitro*- matured oocytes have been successfully used but greater interest has been shown on development of more efficient *in vitro* oocyte maturation system for livestock oocytes, as it is believed to provide an abundant and stable supply of recipient oocytes from slaughter animals. It has also been proposed that oocytes derived from prepubertal animals have reduced developmental competence compared to oocytes from adult animals. The epigenetic reprogramming also depends on donor cell type and its stage of growth. It remains still unclear as to which is the best donor cell from among all the cell types used for cloning. It is difficult to show significant differences in rate for development into cloned animals among different cell types because of the overall low cloning efficiency. The ability to support development of cloned embryos differs among donor cell lines, even if they are derived from the same tissue or organ, making the comparisons further difficult. These differences among cell lines may be due to epigenetic effects, because even within a primary cell culture, generation of cell lines from the same culture shows that some lines are more suitable than others as donors for cloning. The modifications that occur during primary cell culture may result in genomes that are either more or less capable of being reprogrammed. It is currently believed that nuclei from undifferentiated cells, like embryonic stem cells, are more amenable to correct reprogramming, thereby leading to normal cloned calf production. Quiescent donor cells arrested at G0/G1 phases of cycle have been found to be more successful donors than the dividing ones. Serum starvation and growth arrest by contact inhibition are two most commonly used methods to synchronize cells in G0/G1 cell cycle stage. Specific cell cycle inhibitors have also been used but they have been found to significantly affect fetal survival to term as well as neonatal survival. The addition of roscovitine, a cyclin dependent kinase 2 inhibitor, to donor cells successfully synchronizes donor cell cycle and increases nuclear reprogramming capacity of the donor cells. The limited life span of somatic cells, unlike embryonic stem cells, also poses a limitation. Bovine fetal fibroblast cells, which are commonly used to make transgenic cattle, have 30–50 population doublings before senescence. It has been proposed that gene targeting requires around 45 doublings, though some researchers have reported transgenic calves from clonally derived transgenic cell line with a capacity of 30 population doublings. Recent evidence has shown that the doubling capacity can vary widely between cell lines and that genetics may play a major role in determining this capacity, illustrating the importance of picking the right cell line to work with. It has been suggested that development of strategies to increase the lifespan of cultured cells would expand the window of opportunity for gene targeting. Addition of L-carnosine to culture medium has been found to increase the lifespan of cells in culture. Agents like superoxide dismutase mimetic MnTMPyP or culture under reduced O_2 tension (2%) may contribute to delay senescence in cultured cells by reducing oxidative damage to DNA. Introduction of TERT into somatic cells has also been suggested to delay senescence. It has been shown to immortalize the cell lines when expressed. The unfortunate thing was that when these TERT-immortalized/ transformed cells were used as donor cells, they were unable to support development of cloned sheep embryos, suggesting that these cells were not completely or normally reprogrammed.

Thus from the foregoing discussion it could be concluded that SCNT provides for the most promising method for production of transgenic livestock, if its efficiency could be improved. The number of studies reporting successful generation of transgenic farm animals by SCNT is rising and commercial companies are adopting this approach to produce their transgenic founders. We are currently also working on production of transgenic buffalo and goats using SCNT approach and using fetal fibroblast cells, mammary epithelial cells and species-specific embryonic stem cells, transfected with the transgene and tested positive for its expression, as the donor nuclei. Our aim is to produce transgenic animals secreting human protein in milk, and for the purpose use using various mammary gland specific promoters to drive the targeted expression at the right time in the right place and in right quantities.

Applications of transgenic animal production

1. Animal pharming- Industrialization of transgenic animals

Animal pharming refers to the process of using transgenic animals to produce human drugs. It is staking its claim as a lucrative world-market. The pharmaceutical proteins are targeted to be expressed in urine, milk, blood, sperm or eggs as well as to grow rejection-resistant organs for transplant. Transgenic animals are also employed to produce monoclonal antibodies used in vaccine development. Transgenic animals are costly to produce but they have high value in terms of therapeutical proteins and drugs they can produce. These drugs/ proteins serve numerous purposes like treatment of cystic fibrosis, haemophilia, oesteoporosis, arthritis, malaria and HIV. It is estimated that the cost of producing one transgenic animal ranges from USD 20,000 to 300,000, and only a small portion of the attempts succeed to produce a transgenic animal. It is also estimated that one transgenic animal can produce, in its lifetime, USD 200 to 300 million worth of pharmaceuticals. Automated gene sequencing, availability of gene sequences in databases and biological advantage of using animals in comparison to bacteria, insects and plants, have combined to make pharming a preferred alternative. The traditional methods of producing recombinant proteins in laboratory cell cultures, transgenic bacteria, yeast or animal cells are inherently disadvantageous in that: i) the cells and bacterial cultures require constant monitoring and sampling; ii) costly expansion; iii) difficult isolation and purification of proteins; iv) smaller quantities of recombinant proteins produced and v) lesser activity due to impaired post-translational modifications in lower organisms and in culture as compared to a large animal bioreactor. In view of these advantages and decreased unit cost per protein when produced in an animal, animal transgenesis has reached to industrialization stage and is presumed to grow unchallenged in the future. The first transgenic animal, a mouse, was produced in 1981 in an effort to determine the genes involved in cancer. In 1985, first transgenic farm animal was produced. It was a sheep named Tracy that expressed high levels of human protein, alpha-1-antitrypsin, useful for treating emphysema in humans. Transgenic fish have been bred for expression of human growth hormone. Another example is the recently developed "enviro pig," expressing Phytase gene in its salivary glands to allow for better utilization of phosphorous in the feed stuff. This may prove to be a part of solution to animal waste issues. Transgenic cows expressing human lactoferrin in milk have been produced, exemplifying an important step towards humanized milk production. Transgenic goat expressing tissue plasminogen activator (TPA) and alpha anti-trypsin (α-AT) have been produced. Rabbit expressing α-Glucosidae used for treatment of Pomp's disease have also been produced. Although a variety of proteins have been produced in mammary gland of transgenic animals, it has been found that not every protein could be expressed at the desired high amounts. This necessitates further improvement in transgenic technology.

2. Agricultural applications

The development of SCNT along with remarkable progress in gene mapping and genome sequencing endeavors in livestock, opened a new set of possibilities for introduction of precise genetic modifications for agricultural applications. The host of possibilities includes progress in areas like milk production, growth rate, carcass composition, reproductive performance and disease resistance. Transgenic calves produced through SCNT and producing higher levels of β-casein and k-casein in milk, suggest that the transgenes do influence milk composition. With genetic modification, calves resistant to bovine spongiform encephalopathy have been produced. With existing techniques it is possible to produce transgenic animals with better milk yield, both in terms of quality and quantity, altered production characteristics, faster growth rates, lesser age of maturation, better fertility, more carcass composition, better and high quality fibers especially wool etc. could be developed. The day is not far when we will be able to produce pashmina in ordinary sheep or goats under normal climate conditions, surpassing both the need of subzero temperature and the special pashmina breed goats.

3. Human disease models

The success of SCNT has provided a means by which the generation of mammalian animal models other than mice is possible. The mouse is the usual animal model of choice due to factors like short generation time, low maintenance and cost and ease of availability. Although livestock, like pigs and cattle, do not have these attributes that one might consider as a prerequisite for an animal model, they do have similarities to humans that make them great animal models for human diseases. For instance, factors such as lifespan, size and possibly genomic organization are all more similar between cattle or pigs and humans than between mice and humans. Although no one would dispute the impact that the mouse has had as a model on understanding human disease, there are diseases like cystic fibrosis in which the mouse does not display all of the human phenotypes of the disease. Sheep is considered as a possible model for cystic fibrosis due to similarities between the lung in sheep and humans. Similarities between humans and livestock may help advance our understanding of certain diseases and thus understand their pathogenesis, response to drugs and therapies which would finally help in their alleviation.

4. Xenotransplantation

Organ tansplantation is currently one of the most fascinating biomedical techniques. There are more than 250,000 people alive only because of transplantation of an appropriate human organ (allotransplantation). The enormous progress in organ transplantation technology, though being the basis for a normal life of thousands of patients, is experiencing an acute shortage of appropriate organs. This has led to the sad and ethically challenging situation as several thousand patients are reported to die every year throughout the world because of unavailability of the appropriate organ.

To close the growing gap between demand and availability of appropriate organs, xenotransplantation is considered as the solution of choice. Xenotransplantation is referred to the transplantation of organs between discordant species, e.g., form animals to humans. The pig seems to be the optimal donor animal because: i) its organs have size similar to human organs; ii) its anatomy and physiology are not different from humans; iii) pigs have short reproduction cycles and large litters; iv) pig grows rapidly; v) their maintenance is possible at high hygienic standards at relatively low costs; vi) pigs being domesticated species further adds to their ease of exploitation for the purpose.

Despite of these advantages, the use of transgenic pigs as potential donors involves a variety of complex steps and is extremely time-, labour- and resource-intensive. Some of the essential pre-requisites are:

1. Prevention of transmission of zoonoses from the donor animal to the human recipient.

2. Compatibility of the donor organs in anatomy and physiology with the human organ system in terms of life span differences, growth rate and expected body weight.

3. Overcoming the immunological rejection of the transplanted organ. The immunological hurdles include hyperacute rejection response (HAR) which occurs within seconds or minutes, acute vascular rejection (AVR) which occurs within days, cellular rejection which occurs within weeks after transplantation and chronic rejection which results in organ rejection after several years.

The pre-eminent goal of xenotransplantation is overcoming the HAR. This cannot be achieved by administration of high doses of the appropriate immunosuppressive drugs as these do not affect the complement regulated rejection process. The most promising strategy is considered to be the synthesis of human complement regulatory proteins in transgenic pigs. The porcine organ is engineered to produce the complement regulatory protein following transplantation which prevents the complement attack of the recipient. Pigs transgenic for decay accelerating factor (DAF) have been generated and their hearts have been transplanted either heterotrophically (in addition to recipient's own organ) or orthotopically (as life supportive) into non-human primates. The average survival reached to 40–90 days upon heterotrophic transplantation in transgenic animals, whereas the organs were destroyed within a few minutes in non-transgenic animals. Another strategy exploited for making pig a successful donor for xenotransplantation is the knockout of the antigenic structures on the surface of porcine organ. These structures are known as 1,3-α-gal-epitopes and are produced from the gene for 1,3- α-galactosyltransferase. This gene is inactive in humans and old World primates, unlike pigs which demands its knock out for successful xenotransplantation. The birth of piglets with disruption of both allelic loci has been achieved and it is hoped that they would be extremely useful for successful transplantation.

Ethics and regulations

The advancement in transgenic technology, have lead to development of concerns regarding the safety of food supplies and pharmaceutical products, genetic pollution with transgenes as well as zoonotic threat. Despite the promises that transgenic technology holds, there has been no specific legislation enacted regarding transgenic animal use. The US Food and Drug Administration (FDA) has considered transgenic animals according to animal drug provisions of the Federal Food, Drug and Cosmetic Act. As some transgenic animals are produced for food rather than for pharmaceuticals, the US Department of Agriculture, Food safety and Inspection Services (USDA-FSIS) has issued guidance on food safety issues related to transgenics. The FSIS regulations require the producer to supply information about the animals, including any drugs, biological or chemicals that were administered as well as genetic alterations. The Animal and Plant Health Inspection Service (APHIS) of the USDA also addresses questions concerning animal health and disease control, as well as vaccines produced by transgenics.

Further reading

Niemann H and Kues WA. (2003). Applications of tramsgenesis in livestock for agriculture and biomedicine. Animal Reproduction Science, 79: 291–317.

Auchincloss Jr and Sachs DH. (1998). Xenogenic transplantation. Annual Reviews in Immunology, 16: 433–470.

Björklund A. (1991). Neural transplantation—an experimental tool with clinical possibilities. Trends in Neurosciences, 14: 319–322.

Cozzi E and White D. (1995). The generation of transgenic pigs as potential organ donors for humans. Natural Medicine, 1: 964–966.

Jiang XR, Jimenez G, Chang E, Frolkis M, Kusler B and Sage M. (1999). Telomerase expression in human somatic cells does not induce changes associated with a transformed phenotype. Natural Genetics, 21:111–114.

Clark AJ, Burl S, Denning C and Dickinson P. (2000). Gene targeting in livestock: a review. Transgenic Research, 9:263–75.

Bosch P, Hodges CA and Stice SL. (2004). Generation of transgenic livestock by somatic cell nuclear transfer. Biotechnologia Aplicada, 21: 128–136.

Smith Kand Spadafora C. (2005). Sperm mediated gene transfer: applications and implications. Bio Essays, 27:551–562.

Van de Lavoir MC, Diamond JH, Leighton PA, Mather-Love C, Heyer BS, Bradshaw R, Kerchner A, Hooi LT, Gessara TM, Swanberg SE, Delany ME and Etches RJ. (2006). Germline transmission of genetically modified primordial germ cells. Nature, 441:766–769.

Yong HY, Hao Y, Lai L, Li R, Murphy CN, Rieke A, Wax D, Samuel M and Prather RS. (2006). Production of a transgenic piglet by a sperm injection technique in which no chemical or physical treatments were used for oocytes or sperm. Molecular Reproduction and Development, 73:595–599.

Shinohara ET, Kaminski JM, Segal DJ, Pelczar P, Kolhe R, Ryan T, Coates CJ, Fraser MJ, Handler AM, Yanagimachi R and Moisyadi. (2007). Active integration: new strategies for transgenesis. Transgenic Research, 16:333–339.

Robl JM, Wang Z, Kasinathan P and Kuroiwa Y. (2007). Transgenic animal production and animal biotechnology. Theriogenology, 67:127–133.

Houdebine LM. (2005). Use of transgenic animals to improve human health and animal production. Reproduction in Domestic Animals, 40:269–281.

Abbott A. (2004). Laboratory animals: the Renaissance rat. Nature, 428:464–466.

Bachmann A and Knust E. (2008). The use of P-element transposons to generate transgenic flies. Methods in Molecular Biology, 420:61–77.

Bushman F. (2004). Gene regulation: selfish elements make a mark. Nature, 429:253–255.

Fan J and Watanabe T. (2000). Transgenic rabbits expressing human apolipoprotein(a). Journal of Atherosclerosis Thrombosis, 7:8–13.

Donovan DM, Kerr DE and Wall RJ. (2005). Engineering disease resistant cattle. Transgenic Research, 14:563–567.

Houdebine LM. (2004). Preparation of recombinant proteins in milk. Methods in Molecular Biology, 267:485–494.

Redwan el RM, (2009). Animal-derived pharmaceutical proteins. Journal of Immunoassay and Immunochemistry, 30:262–290.

Sosa GM, Gasperi R and Elder GA. (2010). Animal transgenesis: an overview. Brain Structure and Function, 214: 91–109.